Marine Biology

Marine Biology

Timothy Garner

R Callisto Reference

www.callistoreference.com

Callisto Reference,
118-35 Queens Blvd., Suite 400,
Forest Hills, NY 11375, USA

Visit us on the World Wide Web at:
www.callistoreference.com

ISBN: 978-1-64116-555-6 (Hardback)

Cataloging-in-Publication Data

Marine biology / Timothy Garner.
    p. cm.
Includes bibliographical references and index.
ISBN 978-1-64116-555-6
1. Marine biology. 2. Aquatic biology. 3. Biology.
4. Marine sciences. I. Garner, Timothy.
QH91 .M37 2022
578.77--dc23

TABLE OF CONTENTS

PREFACE

The purpose of this book is to help students understand the fundamental concepts of this discipline. It is designed to motivate students to learn and prosper. I am grateful for the support of my colleagues. I would also like to acknowledge the encouragement of my family.

The scientific study of organisms in the sea is referred to as marine biology. It classifies the species on the basis of the environment rather than on taxonomy. There are various habitats which are studied in marine biology such as kelp forests, coral reefs, thermal vents, seagrass meadows and tidepools. There are a wide range of organisms that are studied in marine biology, from zooplankton and phytoplankton to cetaceans. There are many sub-fields of marine biology. A few of them are ichthyology, phycology, invertebrate zoology, etc. Marine biology also studies the physical effects of continual immersion in sea water, adaptation to a salty environment, as well as the effects of various changing oceanic properties on marine life. This textbook attempts to understand the multiple branches that fall under the discipline of marine biology and how such concepts have practical applications. It aims to equip students and experts with the advanced topics and upcoming concepts in this area. This book is an essential guide for both academicians and those who wish to pursue this discipline further.

A foreword for all the chapters is provided below:

Chapter – The Ocean Enviroment

Oceans are the large water bodies that make much of the Earth's hydrosphere. They play an essential role in regulating the Earth's climate. Some of the resources which are obtained from the ocean are fishes, oxygen, diamonds, silver and metal ores. This is an introductory chapter which will introduce briefly all these significant resources which are obtained from the oceans.

Chapter – Marine Organisms

Marine organisms include plants, animals and other organisms that live in the oceans. They also include communities of organisms like Benthos, Nekton, Plankton, etc. The chapter closely examines these types of marine organisms as well as their subtypes to provide an extensive understanding of the subject.

Chapter – Marine Ecosystem

Marine ecosystem is the Earth's largest aquatic ecosystem and is characterized by a high salt content. Some of the areas of study closely associated with marine ecosystems are marine habitats, coral reefs and intertidal zones. This chapter has been carefully written to provide an easy understanding of these facets of marine ecosystem.

Chapter – Threats to Marine Ecosystem

There are numerous factors which threaten the health of marine ecosystems. Ocean dumping, land runoff, ocean acidification, pollution are some of the common threats to marine ecosystems. The topics elaborated in this will help in gaining a better perspective about these threats to marine ecosystems as well as the conservation practices.

Chapter – Relationship of Humans and Oceans

Oceans are the lifeline for human survival as majority of oxygen is produced by the sea plants. They also play a vital role in improving the mental, psychological and emotional wellbeing of humans. The chapter closely examines the relation between human beings and the sea as well as the effect of ocean pollution on humans.

Timothy Garner

The Ocean Environment 1

- **Ocean**
- **Ocean Resources**
- **Future Ocean Resources: Metal-rich Minerals and Genetics**

Oceans are the large water bodies that make much of the Earth's hydrosphere. They play an essential role in regulating the Earth's climate. Some of the resources which are obtained from the ocean are fishes, oxygen, diamonds, silver and metal ores. This is an introductory chapter which will introduce briefly all these significant resources which are obtained from the oceans.

Earth is often referred to as the Water Planet. It is the only planet in our solar system known to have living oceans that are home to marine life. The oceans have been evolving for over 3 billion years. It is only very recently, within the last few 100 years, that any single species has had any significant negative effect on the natural balance of the ocean's ecosystem. Earth's human population has now increased to over 6 billion with over 3 billion now living within 60 miles of a coast line. With the total world population projected to reach 9 billion by 2050, an additional 40 million people are likely to be added to these coastal areas each year up to that date.

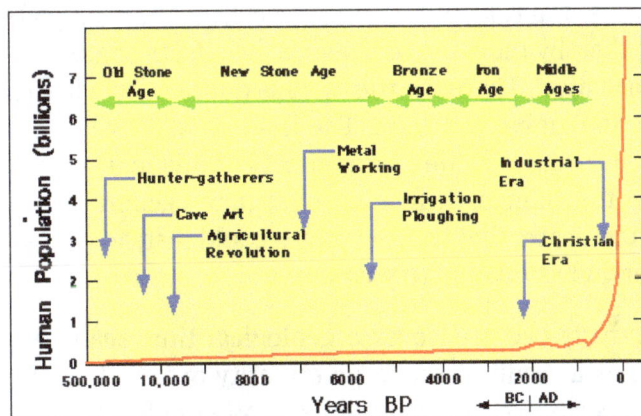

Pollution, over-harvesting and general habitat destruction are seriously impacting our vast, yet extremely fragile, marine environment.

The ocean is the birth place of life on earth. Over several billions of years, it has developed into what it is today through a very intricate and complex balance of natural phenomena. The ocean remains home to several hundred thousand different plant and animal species. It is essential to all living beings, both in the water and on land. The oceans provide the most basic needs such as the oxygen we need to breathe and much of the food we eat. More recently, the ocean has been discovered to provide lifesaving medicines that fight cancer and countless other ailments.

Seals, turtles, fish, and other marine life frequently become entangled in drifting nets and other trash.

Oil covered bird.

The oceans also play an essential role in the carbon cycle, and currently absorb about half of all of the atmospheric carbon, thereby reducing or slowing the effects of global warming. See our Ocean Facts page for more information. There is little serious doubt as to whether or not we humans are causing ozone depletion and global warming. The issue is what will occur due to these changes. To date, most debates concerning the long term effects of this man-made increase in greenhouse gases have assumed that the changes will occur over many generations. However, there is now alarming evidence which indicates that these changes could instead cause an abrupt change in the global climate within a period as little as 10 years.

In the past, fluctuations have occurred over a geological time scale. Now, changes may occur during a human lifetime as a result of the large quantity of heat-trapping gasses released from human activities. Fossil records, ice core samples, and other paleoclimatic data shows that abrupt changes have occurred in the ocean's currents. In turn, these have caused rapid changes to the global temperatures and weather patterns. The ocean makes earth's weather and regulates its temperature while storing over 1,000 times more heat than our atmosphere. The ocean conveyor currents redistribute this huge amount of heat around the planet. This causes hurricanes, typhoons, severe or mild winters, monsoon seasons, El Niños, rainfall patterns, and other climate fluctuations. Small changes in temperature and salinity can and have, as records now show, cause these great currents, like the Gulf Stream, to stop or change course. A 2002 report by the US National Academy of Sciences (NAS) said "available evidence suggests that abrupt climate changes are not only possible but likely in the future, potentially with large impacts on ecosystems and societies."

Additionally, irresponsible actions taken by the exploding human population has caused the total extinction of several hundred plant and animal species during the last few hundred years. According to information published by the World Conservation Union (IUCN) in collaboration with over 600 scientists, 25% of mammal and amphibian species, 11% of birds, 20% of reptiles, and 34% of fish species surveyed so far, are threatened with extinction. In addition, another 5-14% of species in these groups are "nearing threatened" status.

Today, some scientists believe that we may be losing over 30,000 plant and animal species a year. This rate is much faster than at any time since the last great natural extinction 65 million years ago that killed most of the dinosaurs. Seven out of 10 biologists believe the world is now in the midst of the fastest mass extinction of living things in the 4.5 billion-year history of the planet according to a poll conducted by the American Museum of Natural History and the Louis Harris survey research firm. This loss of plant and animal species will also result in the the loss of any potential benefits these species may have provided to us. The possibilities include discovering new medicines and potential cures to human ailments. About 25% of drugs prescribed in the U.S. include chemical compounds derived from wild organisms. Billions of people worldwide rely on plant- and animal-based traditional medicine for their primary health care.

In 1998 at the start of the United Nations International Year of the Ocean, more than 1,600 marine scientists and conservation biologists from 65 countries issued an unprecedented warning to the world's governments and citizens that the sea is in trouble. The consensus is that destruction of marine biological diversity stems from five primary causes:

1. Overexploitation of species,

2. Physical alteration of ecosystems,

3. Pollution,

4. Alien species from distant waters disrupting local food webs,

5. Global atmospheric change.

These findings were again identified more recently in an in-depth study and evaluation of the oceans surrounding the United States. The study by the Pew Oceans Commission is entitled "America's Living Oceans: Charting a Course for Sea Change". The Pew Oceans Commission, chaired by Leon E. Panetta, added Aquaculture and By-catch to their list of identified serious threats. The following pages will explain each of these identified threats in depth and, most importantly, show you what you can do.

The problems that have occurred, and will continue to occur as the delicate balance of ecosystems are interrupted, are only beginning to be understood. We do know that the effects are now well documented on a global scale.

Ocean

Oceanic Divisions

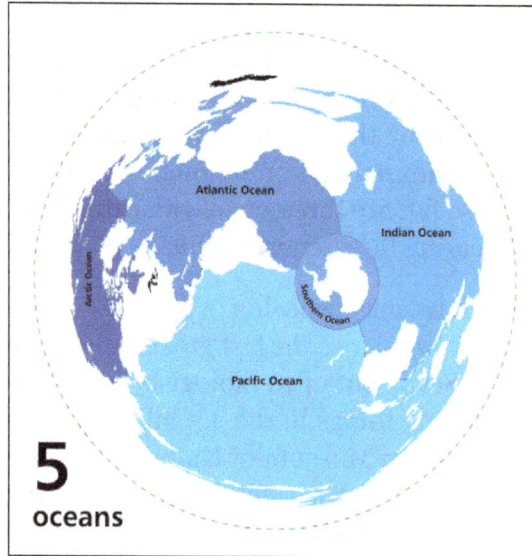

Various ways to divide the World Ocean.

Epipelagic zone: surface – 200 meters deep 2. Mesopelagic zone: 200 m – 1000 m 3. Bathypelagic zone: 1000 m – 4000 m 4. Abyssopelagic zone: 4000 m – 6000 m 5. Hadal zone (the trenches): 6000 m to the bottom of the ocean.

Though generally described as several separate oceans, the global, interconnected body of salt water is sometimes referred to as the World Ocean or global ocean. The concept of a continuous body of water with relatively free interchange among its parts is of fundamental importance to oceanography.

The major oceanic divisions – listed below in descending order of area and volume – are defined in part by the continents, various archipelagos, and other criteria.

	Ocean	Location	Area (km²) (%)	Volume (km³) (%)	Avg. depth (m)	Coastline (km)
1	Pacific Ocean	Separates Asia and Oceania from the Americas.	168,723,000 *46.6*	669,880,000 *50.1*	3,970	135,663
2	Atlantic Ocean	Separates the Americas from Europe and Africa.	85,133,000 *23.5*	310,410,900 *23.3*	3,646	111,866
3	Indian Ocean	Washes upon southern Asia and separates Africa and Australia.	70,560,000 *19.5*	264,000,000 *19.8*	3,741	66,526
4	Southern Ocean	Sometimes considered an extension of the Pacific, Atlantic and Indian Oceans, which encircles Antarctica.	21,960,000 *6.1*	71,800,000 *5.4*	3,270	17,968
5	Arctic Ocean	Sometimes considered a sea or estuary of the Atlantic, which covers much of the Arctic and washes upon northern North America and Eurasia.	15,558,000 *4.3*	18,750,000 *1.4*	1,205	45,389
	Total – World Ocean		361,900,000 *100*	$1.335×10^9$ *100*	3,688	377,412

Seas and Bays			
Sea	Location	Area (sq. km)	#
Arabian Sea	Between the Arabian peninsula and the Indian subcontinent	3,862,000	1
Bay of Bengal	Between the Indian subcontinent and the malaysia peninsula	2,173,000	2

Global System

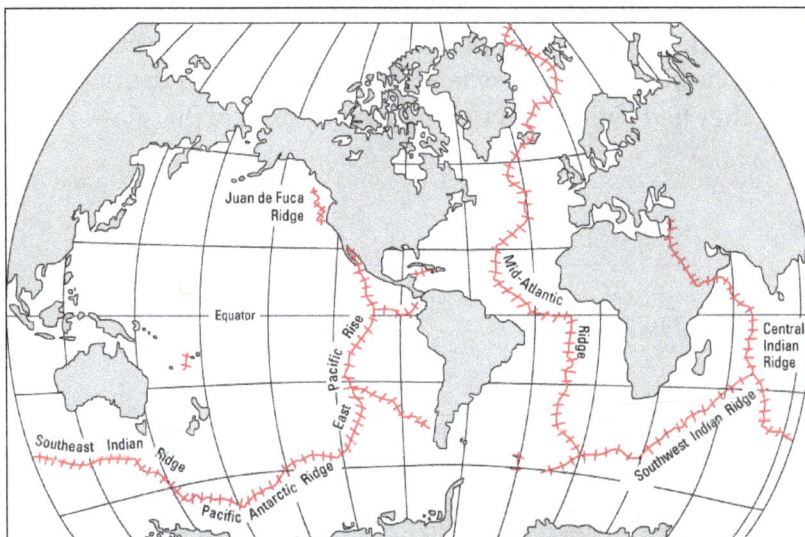

World Distribution of Mid-Oceanic Ridges; USGS.

The mid-ocean ridges of the world are connected and form a single global mid-oceanic ridge system that is part of every ocean and the longest mountain range in the world. The continuous mountain range is 65,000 km (40,000 mi) long (several times longer than the Andes, the longest continental mountain range).

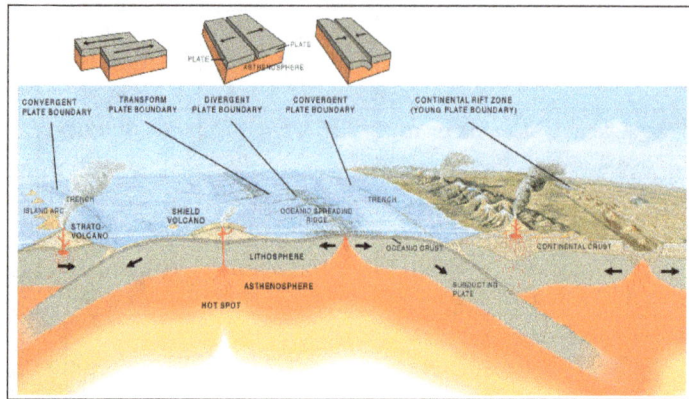

Three main types of plate boundaries.

Physical Properties

The total mass of the hydrosphere is about 1.4 quintillion tonnes (1.4×10^{18} long tons or 1.5×10^{18} short tons), which is about 0.023% of Earth's total mass. Less than 3% is freshwater; the rest is saltwater, almost all of which is in the ocean. The area of the World Ocean is about 361.9 million square kilometers (139.7 million square miles), which covers about 70.9% of Earth's surface, and its volume is approximately 1.335 billion cubic kilometers (320.3 million cubic miles). This can be thought of as a cube of water with an edge length of 1,101 kilometers (684 mi). Its average depth is about 3,688 meters (12,100 ft), and its maximum depth is 10,994 meters (6.831 mi) at the Mariana Trench. Nearly half of the world's marine waters are over 3,000 meters (9,800 ft) deep. The vast expanses of deep ocean (anything below 200 meters or 660 feet) cover about 66% of Earth's surface. This does not include seas not connected to the World Ocean, such as the Caspian Sea.

The bluish ocean color is a composite of several contributing agents. Prominent contributors include dissolved organic matter and chlorophyll. Mariners and other seafarers have reported that the ocean often emits a visible glow which extends for miles at night. In 2005, scientists announced that for the first time, they had obtained photographic evidence of this glow. It is most likely caused by bioluminescence.

Oceanic Zones

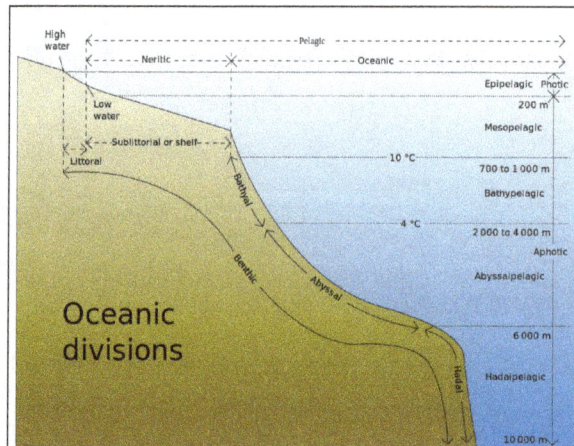

The major oceanic zones, based on depth and biophysical conditions.

Oceanographers divide the ocean into different vertical zones defined by physical and biological conditions. The pelagic zone includes all open ocean regions, and can be divided into further regions categorized by depth and light abundance. The photic zone includes the oceans from the surface to a depth of 200 m; it is the region where photosynthesis can occur and is, therefore, the most biodiverse. Because plants require photosynthesis, life found deeper than the photic zone must either rely on material sinking from above or find another energy source. Hydrothermal vents are the primary source of energy in what is known as the aphotic zone (depths exceeding 200 m). The pelagic part of the photic zone is known as the epipelagic.

The pelagic part of the aphotic zone can be further divided into vertical regions according to temperature. The mesopelagic is the uppermost region. Its lowermost boundary is at a thermocline of 12 °C (54 °F), which, in the tropics generally lies at 700–1,000 meters (2,300–3,300 ft). Next is the bathypelagic lying between 10 and 4 °C (50 and 39 °F), typically between 700–1,000 meters (2,300–3,300 ft) and 2,000–4,000 meters (6,600–13,100 ft), lying along the top of the abyssal plain is the abyssopelagic, whose lower boundary lies at about 6,000 meters (20,000 ft). The last zone includes the deep oceanic trench, and is known as the hadalpelagic. This lies between 6,000–11,000 meters (20,000–36,000 ft) and is the deepest oceanic zone.

The benthic zones are aphotic and correspond to the three deepest zones of the deep-sea. The bathyal zone covers the continental slope down to about 4,000 meters (13,000 ft). The abyssal zone covers the abyssal plains between 4,000 and 6,000 m. Lastly, the hadal zone corresponds to the hadalpelagic zone, which is found in oceanic trenches.

The pelagic zone can be further subdivided into two subregions: the neritic zone and the oceanic zone. The neritic zone encompasses the water mass directly above the continental shelves whereas the oceanic zone includes all the completely open water.

In contrast, the littoral zone covers the region between low and high tide and represents the transitional area between marine and terrestrial conditions. It is also known as the intertidal zone because it is the area where tide level affects the conditions of the region.

If a zone undergoes dramatic changes in temperature with depth, it contains a thermocline. The tropical thermocline is typically deeper than the thermocline at higher latitudes. Polar waters, which receive relatively little solar energy, are not stratified by temperature and generally lack a thermocline because surface water at polar latitudes are nearly as cold as water at greater depths. Below the thermocline, water is very cold, ranging from −1 °C to 3 °C. Because this deep and cold layer contains the bulk of ocean water, the average temperature of the world ocean is 3.9 °C. If a zone undergoes dramatic changes in salinity with depth, it contains a halocline. If a zone undergoes a strong, vertical chemistry gradient with depth, it contains a chemocline.

The halocline often coincides with the thermocline, and the combination produces a pronounced pycnocline.

Exploration

The deepest point in the ocean is the Mariana Trench, located in the Pacific Ocean near the Northern Mariana Islands. Its maximum depth has been estimated to be 10,971 meters (35,994 ft) The

British naval vessel *Challenger II* surveyed the trench in 1951 and named the deepest part of the trench the "Challenger Deep". In 1960, the Trieste successfully reached the bottom of the trench, manned by a crew of two men.

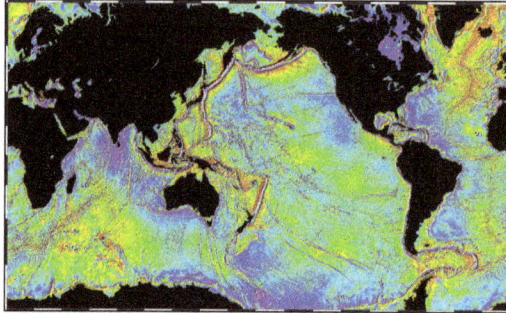

Map of large underwater features (1995, NOAA).

Oceanic Maritime Currents

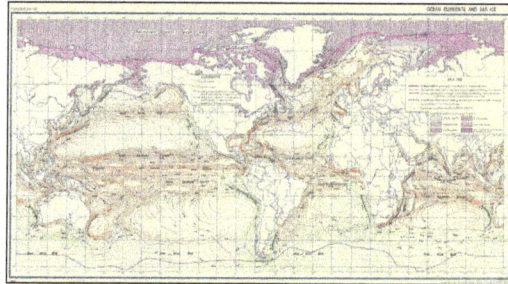

Oceanic surface currents (U.S. Army, 1943).

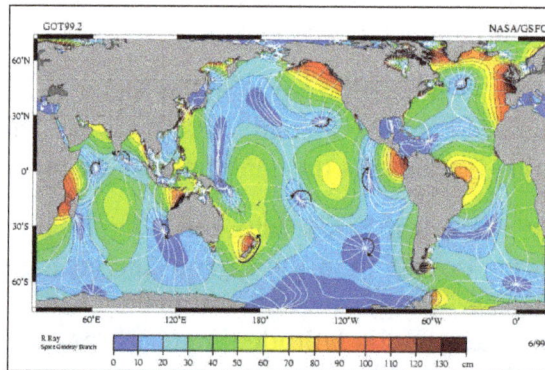

Amphidromic points showing the direction of tides per incrementation periods along with resonating directions of wavelength movements.

Oceanic maritime currents have different origins. Tidal currents are in phase with the tide, hence are quasiperiodic; they may form various knots in certain places, most notably around headlands. Non-periodic currents have for origin the waves, wind and different densities.

The wind and waves create surface currents (designated as "drift currents"). These currents can decompose in one quasi-permanent current (which varies within the hourly scale) and one movement of Stokes drift under the effect of rapid waves movement (at the echelon of a couple of seconds).). The quasi-permanent current is accelerated by the breaking of waves, and in a lesser governing effect, by the friction of the wind on the surface.

This acceleration of the current takes place in the direction of waves and dominant wind. Accordingly, when the sea depth increases, the rotation of the earth changes the direction of currents in proportion with the increase of depth, while friction lowers their speed. At a certain sea depth, the current changes direction and is seen inverted in the opposite direction with current speed becoming null: known as the Ekman spiral. The influence of these currents is mainly experienced at the mixed layer of the ocean surface, often from 400 to 800 meters of maximum depth. These currents can considerably alter, change and are dependent on the various yearly seasons. If the mixed layer is less thick (10 to 20 meters), the quasi-permanent current at the surface adopts an extreme oblique direction in relation to the direction of the wind, becoming virtually homogeneous, until the Thermocline.

In the deep however, maritime currents are caused by the temperature gradients and the salinity between water density masses.

In littoral zones, breaking waves are so intense and the depth measurement so low, that maritime currents reach often 1 to 2 knots.

Climate

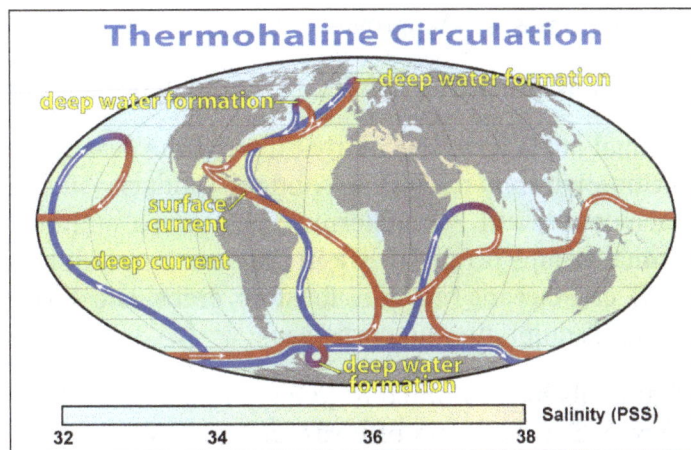

A map of the global thermohaline circulation; blue represent deep-water currents, whereas red represent surface currents.

Ocean currents greatly affect Earth's climate by transferring heat from the tropics to the polar regions. Transferring warm or cold air and precipitation to coastal regions, winds may carry them inland. Surface heat and freshwater fluxes create global density gradients that drive the thermohaline circulation part of large-scale ocean circulation. It plays an important role in supplying heat to the polar regions, and thus in sea ice regulation. Changes in the thermohaline circulation are thought to have significant impacts on Earth's energy budget. In so far as the thermohaline circulation governs the rate at which deep waters reach the surface, it may also significantly influence atmospheric carbon dioxide concentrations.

The Antarctic Circumpolar Current encircles that continent, influencing the area's climate and connecting currents in several oceans.

One of the most dramatic forms of weather occurs over the oceans: tropical cyclones (also called "typhoons" and "hurricanes" depending upon where the system forms).

Biology

The ocean has a significant effect on the biosphere. Oceanic evaporation, as a phase of the water cycle, is the source of most rainfall, and ocean temperatures determine climate and wind patterns that affect life on land. Life within the ocean evolved 3 billion years prior to life on land. Both the depth and the distance from shore strongly influence the biodiversity of the plants and animals present in each region.

As it is thought that life evolved in the ocean, the diversity of life is immense, including:

- Bacteria: Ubiquitous single-celled prokaryotes found throughout the world.

- Archaea: Prokaryotes distinct from bacteria, that inhabit many environments of the ocean, as well as many extreme environments.

- Algae: Algae is a "catch-all" term to include many photosynthetic, single-celled eukaryotes, such as green algae, diatoms, and dinoflagellates, but also multicellular algae, such as some red algae (including organisms like Pyropia, which is the source of the edible nori seaweed), and brown algae (including organisms like kelp).

- Plants: Including sea grasses, or mangroves.

- Fungi: Many marine fungi with diverse roles are found in oceanic environments.

- Animals: Most animal phyla have species that inhabit the ocean, including many that are only found in marine environments such as sponges, Cnidaria (such as corals and jellyfish), comb jellies, Brachiopods, and Echinoderms (such as sea urchins and sea stars). Many other familiar animal groups primarily live in the ocean, including cephalopods (includes octopus and squid), crustaceans (includes lobsters, crabs, and shrimp), fish, sharks, cetaceans (includes whales, dolphins, and porpoises).

In addition, many land animals have adapted to living a major part of their life on the oceans. For instance, seabirds are a diverse group of birds that have adapted to a life mainly on the oceans. They feed on marine animals and spend most of their lifetime on water, many only going on land for breeding. Other birds that have adapted to oceans as their living space are penguins, seagulls and pelicans. Seven species of turtles, the sea turtles, also spend most of their time in the oceans.

Salinity

A zone of rapid salinity increase with depth is called a halocline. The temperature of maximum density of seawater decreases as its salt content increases. Freezing temperature of water decreases with salinity, and boiling temperature of water increases with salinity. Typical seawater freezes at around −2 °C at atmospheric pressure. If precipitation exceeds evaporation, as is the case in polar and temperate regions, salinity will be lower. If evaporation exceeds precipitation, as is the case in tropical regions, salinity will be higher. Thus, oceanic waters in polar regions have lower salinity content than oceanic waters in temperate and tropical regions.

Salinity can be calculated using the chlorinity, which is a measure of the total mass of halogen ions (includes fluorine, chlorine, bromine, and iodine) in seawater. By international agreement, the following formula is used to determine salinity:

Salinity (in ‰) = 1.80655 × Chlorinity (in ‰).

The average chlorinity is about 19.2‰, and, thus, the average salinity is around 34.7‰.

Absorption of Light

Absorption of light in different wavelengths by ocean		
Color: Wavelength (nm)	Depth at which 99 percent of the wavelength is absorbed (in meters)	Percent absorbed in 1 meter of water
Ultraviolet (UV): 310	31	14.0
Violet (V): 400	107	4.2
Blue (B): 475	254	1.8
Green (G): 525	113	4.0
Yellow (Y): 575	51	8.7
Orange (O): 600	25	16.7
Red (R): 725	4	71.0
Infrared (IR): 800	3	82.0

Economic Value

Many of the world's goods are moved by ship between the world's seaports. Oceans are also the major supply source for the fishing industry. Some of the major harvests are shrimp, fish, crabs, and lobster.

Waves and Swell

The motions of the ocean surface, known as undulations or *waves*, are the partial and alternate rising and falling of the ocean surface. The series of mechanical waves that propagate along the interface between water and air is called swell.

Extraterrestrial Oceans

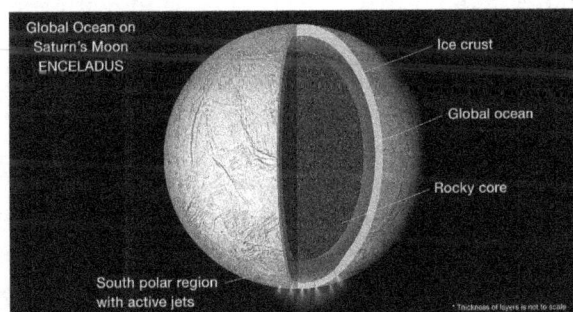

Artist's conception of subsurface ocean of Enceladus.

Although Earth is the only known planet with large stable bodies of liquid water on its surface and the only one in the Solar System, other celestial bodies are thought to have large oceans.

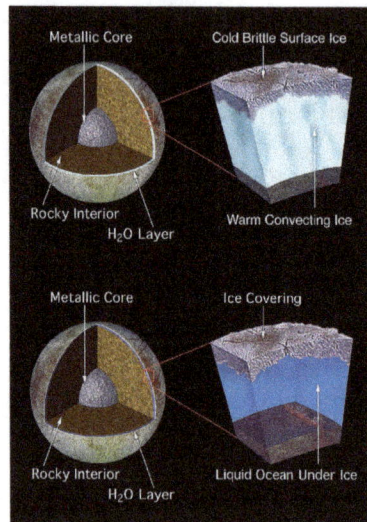

Two models for the composition of Europa predict a large subsurface ocean of liquid water.
Similar models have been proposed for other celestial bodies in the Solar System.

Planets

The gas giants, Jupiter and Saturn, are thought to lack surfaces and instead have a stratum of liquid hydrogen; however their planetary geology is not well understood. The possibility of the ice giants Uranus and Neptune having hot, highly compressed, supercritical water under their thick atmospheres has been hypothesised. Although their composition is still not fully understood, a 2006 study by Wiktorowicz and Ingersall ruled out the possibility of such a water "ocean" existing on Neptune, though some studies have suggested that exotic oceans of liquid diamond are possible.

The Mars ocean hypothesis suggests that nearly a third of the surface of Mars was once covered by water, though the water on Mars is no longer oceanic (much of it residing in the ice caps). The possibility continues to be studied along with reasons for their apparent disappearance. Astronomers now think that Venus may have had liquid water and perhaps oceans for over 2 billion years.

Natural Satellites

A global layer of liquid water thick enough to decouple the crust from the mantle is thought to be present on the natural satellites Titan, Europa, Enceladus and, with less certainty, Callisto, Ganymede and Triton. A magma ocean is thought to be present on Io. Geysers have been found on Saturn's moon Enceladus, possibly originating from an ocean about 10 kilometers (6.2 mi) beneath the surface ice shell. Other icy moons may also have internal oceans, or may once have had internal oceans that have now frozen.

Large bodies of liquid hydrocarbons are thought to be present on the surface of Titan, although they are not large enough to be considered oceans and are sometimes referred to as *lakes* or seas. The Cassini–Huygens space mission initially discovered only what appeared to be dry lakebeds and empty river channels, suggesting that Titan had lost what surface liquids it might have had. Later flybys of Titan provided radar and infrared images that showed a series of hydrocarbon lakes in the colder polar regions. Titan is thought to have a subsurface liquid-water ocean under the ice in addition to the hydrocarbon mix that forms atop its outer crust.

Dwarf Planets and Trans-Neptunian Objects

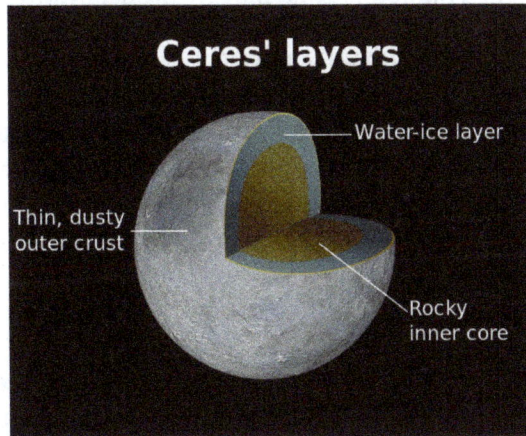

Diagram showing a possible internal structure of Ceres.

Ceres appears to be differentiated into a rocky core and icy mantle and may harbour a liquid-water ocean under its surface.

Not enough is known of the larger trans-Neptunian objects to determine whether they are differentiated bodies capable of supporting oceans, although models of radioactive decay suggest that Pluto, Eris, Sedna, and Orcus have oceans beneath solid icy crusts approximately 100 to 180 km thick.

Extrasolar

Rendering of a hypothetical large extrasolar moon with surface liquid-water oceans.

Some planets and natural satellites outside the Solar System are likely to have oceans, including possible water ocean planets similar to Earth in the habitable zone or "liquid-water belt". The detection of oceans, even through the spectroscopy method, however is likely extremely difficult and inconclusive.

Theoretical models have been used to predict with high probability that GJ 1214 b, detected by transit, is composed of exotic form of ice VII, making up 75% of its mass, making it an ocean planet.

Other possible candidates are merely speculated based on their mass and position in the habitable zone include planet though little is actually known of their composition. Some scientists speculate Kepler-22b may be an "ocean-like" planet. Models have been proposed for Gliese 581 d that could include surface oceans. Gliese 436 b is speculated to have an ocean of "hot ice". Exomoons orbiting planets, particularly gas giants within their parent star's habitable zone may theoretically have surface oceans.

Terrestrial planets will acquire water during their accretion, some of which will be buried in the magma ocean but most of it will go into a steam atmosphere, and when the atmosphere cools it will collapse on to the surface forming an ocean. There will also be outgassing of water from the mantle as the magma solidifies—this will happen even for planets with a low percentage of their mass composed of water, so "super-Earth exoplanets may be expected to commonly produce water oceans within tens to hundreds of millions of years of their last major accretionary impact."

Non-water Surface Liquids

Oceans, seas, lakes and other bodies of liquids can be composed of liquids other than water, for example the hydrocarbon lakes on Titan. The possibility of seas of nitrogen on Triton was also considered but ruled out. There is evidence that the icy surfaces of the moons Ganymede, Callisto, Europa, Titan and Enceladus are shells floating on oceans of very dense liquid water or water–ammonia. Earth is often called *the* ocean planet because it is 70% covered in water. Extrasolar terrestrial planets that are extremely close to their parent star will be tidally locked and so one half of the planet will be a magma ocean. It is also possible that terrestrial planets had magma oceans at some point during their formation as a result of giant impacts. Hot Neptunes close to their star could lose their atmospheres via hydrodynamic escape, leaving behind their cores with various liquids on the surface. Where there are suitable temperatures and pressures, volatile chemicals that might exist as liquids in abundant quantities on planets include ammonia, argon, carbon disulfide, ethane, hydrazine, hydrogen, hydrogen cyanide, hydrogen sulfide, methane, neon, nitrogen, nitric oxide, phosphine, silane, sulfuric acid, and water.

Supercritical fluids, although not liquids, do share various properties with liquids. Underneath the thick atmospheres of the planets Uranus and Neptune, it is expected that these planets are composed of oceans of hot high-density fluid mixtures of water, ammonia and other volatiles. The gaseous outer layers of Jupiter and Saturn transition smoothly into oceans of supercritical hydrogen. The atmosphere of Venus is 96.5% carbon dioxide, which is a supercritical fluid at its surface.

Why is the Ocean so Important?

Seven reasons why the ocean is important:

Ocean Produces more Oxygen than the Amazones

It is often thought that rainforests are the primary source of oxygen on the planet, but the truth is that rainforests are only responsible for 28% of the oxygen on earth while oceans are responsible for the 70%. It does not matter how far we live from, from every ten breaths you take seven come from the ocean.

Have you ever seen a tree in the middle of the sea? No, right? That is because the Ocean does not need them, the phytoplankton has got it covered. Phytoplankton is a microscopic plant, a component of the plankton, which spends its life being carried by oceanic currents. Basically, these tiny little organisms act in the same way as tree leaves do on land. Phytoplankton absorbs carbon dioxide and releases oxygen. We do not see them, so we tend to forget about them if we even know about them in the first place. They are one of the tiniest beings on the planet, but one of the most important to have around, keeping us alive.

The Ocean Regulates the Earth Climate

In many ways, the sea regulates our climate. It soaks up the heat and transports warm water from the equator to the poles, and cold water from the poles to the tropics. Without these currents, the weather would be extreme in some regions, and fewer places would be habitable.

It regulates rain and droughts. Holding 97% of the water of our planet, almost all rain that drops on land comes from the sea. The ocean absorbs CO_2, to keep the carbon cycle, and accordingly temperatures on earth, in balance. It is like our global climate control system.

It is an Important Source of Food

The ocean is the number one source of protein for more than a billion people. Fish accounts for about 15.7% of the animal protein consumed globally. Although, not everything is fish and seafood. Humans have traditionally used algae and sea plants for cooking sushi, seaweed pancit in Philippines, sea grapes, dulse, etc. There is a growing tendency of using algae and sea plants on our daily and start-ups like, "This is seaweed " are making sure to introduce it in our supermarkets.

Considering the world population growing by 1.5 million people every week, we are relying on the ocean more and more for survival, and we need from an alternative and nutritious food sources. For those who are not into eating insects, edible seaweed might be a good alternative.

Many Creatures Depend and Live in the Ocean

The ocean is not just home to us ocean lovers, but it is home to the greatest abundance of life on our planet. When you sail across an ocean, you will see dolphins, whales or a turtle popping up to take a breath. That is just what we see on the surface; there is more life below the ocean's surface than on land. Experts predict that there are more than 300.000 different species underwater, and is still not clear how many of them we know.

All the creatures that live in the Ocean play an essential role in the trophic chain of the ecosystems. Due to climate change, the ocean has been warming and losing oxygen, and its pH has been decreasing. Many marine species have already been adjusting their geographic and depth ranges with changes in sea temperature. However, not all species may be able to move to avoid thermal stress, and global warming has already been correlated with mass mortalities in the Mediterranean.

With more than 60% of the world's population living on the coastline, we all depend on a healthy sea just as much as these beautiful creatures.

Best Holidays are Close to the Water

The ocean is a happy-zone. Our temple, our life, our second home, our exhilaration place. It's where we swim, surf, sail, dive, chillax, and 'lime'. Family holidays and Sundays often happen on the beach. For sailors, fisherman and islanders, it also is a transport zone. It carries us to new lands, and connect us with nature and each other.

As sailors, we also serve as educators, ambassadors and advocates of a lifestyle on the water. Together we share a passion for the ocean, and an avid desire to keep our playground clean and safe forever. Waterways are crucial to our health, for us and future generations.

Many Jobs are Related to Sea Activities

The FAO estimates on the most recent official statistics indicate that 59.6 million people in the world were engaged in fisheries and aquaculture in 2016.

Only at the European Union level, the blue sector represents 3.362.510 of jobs, in 9 subsectors: coastal and maritime tourism, Aquaculture, Renewable Energy, Mineral Resources, Biotechnology, Fisheries, Shipbuilding and Ship repair, Offshore Oil and Gas, and Transport. In fact, 90% of the world trade is made by sea. In the U.S. close to three million jobs are directly dependent on the resources of the oceans and Great Lakes.

The ocean gives jobs to fishers, lifeguards, surf instructors, harbours, (free)diving schools, marine-based tour operators, water sports businesses, holiday accommodations, and, of course, ocean nomads.

Ocean has Therapeutic Properties

Did you know that the anti-viral drugs Zovirax and Acyclovir were obtained from nucleosides isolated from Caribbean sponges? Or that Yondelis, developed from small soft-bodied marine animals, was the first drug of marine origin to fight cancer?

When we dip in the water, our inner dolphin gets released. It's called the "mammalian diving reflex". We learned this when our started freediving. When our face touches water, our heart rate immediately slows down, and blood moves from the extremities to the brain, heart and vital organs of our body. Seals and dolphins have this reflex, and so do we! It wakes us up and makes us feel vibrant and alive.

The ocean is therapeutic. When we see, feel, hear, smell or taste water we are happy and at peace. Research has proven that the so-called blue spaces can directly reduce psychological stress and improve mood. Read Blue Mind to learn more about that.

Despite all that, we still know more about Mars than we know about the ocean! A healthy ocean keeps us healthy on earth. We are alive right now because of the oceans. Now the ocean needs to be kept alive by us. The choices we make now determine our future, and our children's future. We have the responsibility to care for the ocean as it cares for us.

Ocean Resource

Fishing Facts

The oceans have been fished for thousands of years and are an integral part of human society. Fish have been important to the world economy for all of these years, starting with the Viking trade of cod and then continuing with fisheries like those found in Lofoten, Europe, Italy, Portugal, Spain and India. Fisheries of today provide about 16% of the total world's protein with higher percentages occurring in developing nations. Fisheries are still enormously important to the economy and well-being of communities.

Fish Market in the Philippines.

The word fisheries refers to all of the fishing activities in the ocean, whether they are to obtain fish for the commercial fishing industry, for recreation or to obtain ornamental fish or fish oil. Fishing activities resulting in fish not used for consumption are called industrial fisheries. Fisheries are usually designated to certain ecoregions like the salmon fishery in Alaska, the Eastern Pacific tuna fishery or the Lofoten island cod fishery. Due to the relative abundance of fish on the continental shelf, fisheries are usually marine and not freshwater.

Although a world total of 86 million tons of fish were captured in 2000, China's fisheries were the most productive, capturing a whopping one third of the total. Other countries producing the most fish were Peru, Japan, the United States, Chile, Indonesia, Russia, India, Thailand, Norway and Iceland- with Peru being the most and Iceland being the least. The number of fish caught varies with the years, but appears to have leveled off at around 88 million tons per year possibly due to overfishing, economics and management practices.

Fish are caught in a variety of ways, including one-man casting nets, huge trawlers, seining, drift-netting, handlining, longlining, gillnetting and diving. The most common species making up the global fisheries are herring, cod, anchovy, flounder, tuna, shrimp, mullet, squid, crab, salmon, lobster, scallops and oyster. Mollusks and crustaceans are also widely sought. The fish that are caught are not always used for food. In fact, about 40% of fish are used for other purposes such as fishmeal to feed fish grown in captivity. For example cod, is used for consumption, but is also frozen for later use. Atlantic herring is used for canning, fishmeal and fish oil. The Atlantic menhaden is used for fishmeal and fish oil and Alaska pollock is consumed, but also used for fish paste to simulate crab. The Pacific cod has recently been used as a substitute for Atlantic cod which has been overfished.

The amount of fish available in the oceans is an ever-changing number due to the effects of both natural causes and human developments. It will be necessary to manage ocean fisheries in the coming years to make sure the number of fish caught never makes it to zero. A lack of fish greatly impacts the economy of communities dependent on the resource, as can be seen in Japan, eastern Canada, New England, Indonesia and Alaska. The anchovy fisheries off the coast of western South America have already collapsed and with numbers dropping violently from 20 million tons to 4 million tons—they may never fully recover. Other collapses include the California sardine industry, the Alaskan king crab industry and the Canadian northern cod industry. In Massachusetts alone, the cod, haddock and yellowtail flounder industries collapsed, causing an economic disaster for the area.

Due to the importance of fishing to the worldwide economy and the need for humans to understand human impacts on the environment, the academic division of fisheries science was developed. Fisheries science includes all aspects of marine biology, in addition to economics and management skills and information. Marine conservation issues like overfishing, sustainable fisheries and management of fisheries are also examined through fisheries science.

In order for there to be plenty of fish in the years ahead, fisheries will have to develop sustainable fisheries and some will have to close. Due to the constant increase in the human population, the oceans have been overfished with a resulting decline of fish crucial to the economy and communities of the world. The control of the world's fisheries is a controversial subject, as they cannot produce enough to satisfy the demand, especially when there aren't enough fish left to breed in healthy ecosystems. Scientists are often in the role of fisheries managers and must regulate the amount of fishing in the oceans, a position not popular with those who have to make a living fishing ever decreasing populations.

The two main questions facing fisheries management are:

1. What is the carrying capacity of the ocean? How many fish are there and how many of which type of fish should be caught to make fisheries sustainable?

2. How should fisheries resources be divided among people?

Fish populate the ocean in patches instead of being spread out throughout the enormous expanse. The photic zone is only 10-30 m deep near the coastline, a place where phytoplankton have enough solar energy to grow in abundance and fish have enough to eat. Most commercial fishing takes place in these coastal waters, as well as estuaries and the slope of the Continental Shelf. High nutrient contents from upwelling, runoff, the regeneration of nutrients and other ecological processes supply fish in these areas with the necessary requirements for life. The blue color of the water near the coastlines is the result of chlorophyll contained in aquatic plant life.

Most fish are only found in very specific habitats. Shrimp are fished in river deltas that bring large amounts of freshwater into the ocean. The areas of highest productivity known as banks are actually where the Continental Shelf extends outward towards the ocean. These include the Georges Bank near Cape Cod, the Grand Banks near Newfoundland and Browns Bank. Areas where the ocean is very shallow also contain many fish and include the middle and southern regions of the North Sea. Coastal upwelling areas can be found off of southwest Africa and off South America's western coast. In the open ocean, tuna and other mobile species like yellowfin can be found in large amounts.

The question of how many fish there are in the ocean is a complicated one but can be simplified using populations of fish instead of individuals. The word "cohort" refers to the year the fish was born and is used to gather population statistics. Cohorts start off as eggs with an extremely high rate of mortality, which declines as the fish gets older. Juvenile fish close to the age where they can be fished are called "recruits". Cohort mortality is tied in with the species of fish due to variances in natural mortality. The biomass of a particular cohort is greatest when fish are rapidly growing and decreases as the fish get older and start to die.

Scientists use theories and models to help determine the number and size of fish populations in the ocean. Production theory is the theory that production will be highest when the number of fish does not overwhelm the environment and there are not too few for genetic diversity of populations. The maximum sustainable yield is produced when the population is of intermediate size. Yield-per-recruit theory is the quest to determine the optimum age for harvesting fish. The theory of recruitment and stock allows scientists to make a guess about the optimum population size to encourage a larger population of recruits. All of the above theories must be flexible enough to allow natural fluctuations in the fish population to occur and still gather significant data; however, the theories are limited when taking into account the effect of humans on the environment and misinformation could result in overfishing of the ocean's resources.

Other factors that must be taken into account are the ecological requirements of individual fish species like predation and nutrition and why fish will often migrate to different areas. Water temperatures also influence the behavior of ecosystems, causing an increase in metabolism and predation or a sort of hibernation. Even the amount of turbulence in the water can affect predator-prey relationships, with more meetings between the two when waters are stirred up. Global warming could have a huge economic impact on the fisheries when fish stocks are forced to move to waters with more tolerable temperatures.

In many countries, commercial fishing has found more temporarily economical ways of catching fish, including gill nets, purse seines, and drift nets. Although fish are trapped efficiently in one day using these fishing practices, the number of fish that are wasted this way has reached 27 million tons per year, not to mention the crucial habitats destroyed that are essential for the regeneration of fish stocks. In addition, marine mammals and birds are also caught in these nets. The wasted fish and marine life is referred to as bycatch, an unfortunate side-effect of unsustainable fishing practices that can turn the ecosystem upside-down and leave huge amounts of dead matter in the water. Other human activities like trawling and dredging of the ocean floor have bulldozed over entire underwater habitats. The oyster habitat has been completely destroyed in many areas from the use of the oyster patent tong and sediment buildup draining from farm runoff.

Shipping

The word "shipping" refers to the activity of moving cargo with ships in between seaports. Wind-powered ships exist, but more often ships are powered by steam turbine plants or diesel engines. Naval ships are usually responsible for transporting most of trade from one country to another and are called merchant navies. The various types of ships include container ships, tankers, crude oil ships, product ships, chemical ships, bulk carriers, cable layers, general cargo ships, offshore supply vessels, dynamically-positioned ships, ferries, gas and car carriers, tugboats, barges and dredgers.

In theory, shipping can have a low impact on the environment. It is safe and profitable for economies around the world. However, serious problems occur with the shipping of oil, dumping of waste water into the ocean, chemical accidents at sea, and the inevitable air and water pollution occuring when modern day engines are used. Ships release air pollutants in the form of sulphur dioxide, nitrogen oxides, carbon dioxide, hydrocarbons and carbon monoxide. Chemicals dumped in the ocean from ships include chemicals from the ship itself, cleaning chemicals for machine parts, and cleaning supplies for living quarters. Large amounts of chemicals are often spilled into the ocean and sewage is not always treated properly or treated at all. Alien species riding in the ballast water of ships arrive in great numbers to crash native ecosystems and garbage is dumped over the side of many vessels. Dangerous industrial waste and harmful substances like halogenated hydrocarbons, water treatment chemicals, and anti-fouling paints are also dumped frequently. Ships and other watercraft with engines disturb the natural environment with loud noises, large waves, frequently striking and killing animals like manatees and dolphins.

Tourism

Tourism is the fastest growing division of the world economy and is responsible for more than 200 million jobs all over the world. In the US alone, tourism resulted in an economic gain of 478 billion dollars. With 700 million people traveling to another country in the year 2000, tourism is in the top five economic contributors to 83% of all countries and the most important economy for 38% of countries. The tourism industry is based on natural resources present in each country and usually negatively affect ecosystems because it is often left unmanaged. However, sustainable tourism can actually promote conservation of the environment.

Dive boat with recreational divers, Key Largo, Florida.

The negative effects of tourism originate from the development of coastal habitats and the annihilation of entire ecosystems like mangroves, coral reefs, wetlands and estuaries. Garbage and sewage generated by visitors can add to the already existing solid waste and garbage disposal issues present in many communities. Often visitors produce more waste than locals, and much of it ends up as untreated sewage dumped in the ocean. The ecosystem must cope with eutrophication, or the loss of oxygen in the water due to excessive algal bloom, as well as disease epidemics. Sewage can be used as reclaimed water to treat lawns so that fertilizers and pesticides do not seep into the ocean.

Other problems with tourism include the overexploitation of local seafood, the destruction of local habitats through careless scuba diving or snorkeling and the dropping of anchors on underwater features. Ecotourism and cultural tourism are a new trend that favors low impact tourism and fosters a respect for local cultures and ecosystems.

Mining

Humans began to mine the ocean floor for diamonds, gold, silver, metal ores like manganese nodules and gravel mines in the 1950's when the company Tidal Diamonds was established by Sam Collins. Diamonds are found in greater number and quality in the ocean than on land, but are much harder to mine. When diamonds are mined, the ocean floor is dredged to bring it up to the boat and sift through the sediment for valuable gems. The process is difficult as sediment is not easy to bring up to the surface, but will probably become a huge industry once technology evolves to solve the logistical problem.

Metal compounds, gravels, sands and gas hydrates are also mined in the ocean. Mining of manganese nodules containing nickel, copper and cobalt began in the 1960's and soon after it was discovered that Papua New Guinea was one of the few places where nodules were located in shallow waters rather than deep waters. Although manganese nodules could be found in shallow waters in significant quantities, the expense of bringing the ore up to the surface proved to be expensive. Sands and gravels are often mined for in the United States and are used to protect beaches and reduce the effects of erosion.

Mining the ocean can be devastating to the natural ecosystems. Dredging of any kind pulls up the ocean floor resulting in widespread destruction of marine animal habitats, as well as wiping out vast numbers of fishes and invertebrates. When the ocean floor is mined, a cloud of sediment rises up in the water, interfering with photosynthetic processes of phytoplankton and other marine life, in addition to introducing previously benign heavy metals into the food chain. As minerals found on land are exploited and used up, mining of the ocean floor will increase.

Climate Buffer

The ocean is an integral component of the world's climate due to its capacity to collect, drive and mix water, heat, and carbon dioxide. The ocean can hold and circulate more water, heat and carbon dioxide than the atmosphere although the components of the Earth's climate are constantly exchanged. Because the ocean can store so much heat, seasons occur later than they would and air above the ocean is warmed. Heat energy stored in the ocean in one season will affect the climate almost an entire season later. The ocean and the atmosphere work together to form complex weather phenomena like the North Atlantic Oscillation and El Niño. The many chemical cycles occurring between the ocean and the atmosphere also influence the climate by controlling the amount of radiation released into ecosystems and our environment.

The atmosphere directly above the ocean does not absorb much heat by itself, so in order for it to warm up, the temperature of the ocean has to rise first. The two other ways for the atmosphere to warm near the ocean are by reflection of light off of the surface of the ocean or by the evaporation of water from the ocean surface. The temperature of the ocean controls the climate in the lower part of the atmosphere, so for most areas of the Earth the ocean temperature is responsible for the air temperature.

The main forms of climate buffering by the ocean are by the transport of heat through ocean currents traveling across huge basins. Areas like the tropics end up being cooled and higher latitudes are warmed by this effect. Air temperatures worldwide are regulated by the circulation of heat by the oceans. The ocean stores heat in the upper two meters of the photic zone. This is possible because seawater has a very high density and specific heat and can store vast quantities of energy in the form of heat. The ocean can then buffer changes in temperature by storing heat and releasing heat. Evaporation cools ocean water which cools the atmosphere. It is most noticeable near the equator and the effect decreases closer to the poles.

Oxygen Production

Gases in the atmosphere like carbon, nitrogen, sulfur and oxygen are dissolved through the water cycle. The gases that are now crucial to all ecosystems and biological processes originally came from the inside layers of the earth during the period when the earth was first formed. The rate of flow for oxygen as well as other gases is controlled by biological processes, especially metabolism of organisms like prokaryotes and bacteria. Prokaryotes have been around since the beginning of the Earth, have evolved to be able to use chemical energy to create organic matter and are capable of both reducing and oxidizing inorganic compounds. Bacteria that can reduce inorganic compounds are anaerobic and those that oxidize inorganic compounds are aerobic. Aerobic bacteria release oxygen as a by-product of photosynthesis.

Approximately two billion years ago, aerobic bacteria began producing oxygen which gradually filled up all of the oxygen reservoirs in the environment. Once these "sinks" were filled, molecular oxygen began to build in the atmosphere, creating an environment favorable for other life to inhabit the Earth. Sinks included reduced iron ions and hydrogen sulfide gas. Evidence of this process can be found in the banded iron formations created when iron minerals were precipitated. The oxygen started to fill the atmosphere up and new bacteria evolved that could use oxygen to oxidize both inorganic and organic compounds. Bacteria that were accustomed to an oxygen-poor atmosphere only survived in anaerobic environments like sewage, swamps, and in the sediments of both marine and freshwater areas.

Phytoplankton account for possibly 90% of the world's oxygen production because water covers about 70% of the Earth and phytoplankton are abundant in the photic zone of the surface layers. Some of the oxygen produced by phytoplankton is absorbed by the ocean, but most flows into the atmosphere where it becomes available for oxygen dependent life forms.

Resources from the Ocean Floor

The one resource originating in marine deposits and also of prime importance economically at present is petroleum. Energy use is basic to a flourishing economy, so that the item has traditionally received much supportive attention both from elected politicians and from other decision makers in the industry and elsewhere. However, in recent decades, it has become obvious that the use of carbon-based fossil fuels involves uninsurable risks. Manageable and insurable risks are largely confined to activities concerning obtaining the resources. The risks with regard to energy waste disposal (mainly carbon dioxide), however, are much larger, involving unwanted climate change, as well as sea-level rise and other associated serious problems.

Future Ocean Resources: Metal-rich Minerals and Genetics

The ocean provides humanity with many resources and is an important source of food, transport, energy, and recreation. However, it remains famously under-explored (only 10-15% of ocean has been mapped at even 100 m resolution). As scientific and technological progress allows us to expand our knowledge new classes of resources are becoming increasingly feasible.

This Royal Society project considers two emerging classes of ocean based resources, metals from the deep ocean floor, and the application of the genetics and chemicals produced by marine life. An ever increasing world population is driving the need for metals for use in electronics and low carbon technologies and novel drugs, not least antibiotics, to treat disease.

Sustainable use of these novel resources could have significant benefits, but involves interaction with a natural environment that is challenging to access and less well understood than that on land. Exploration for minerals and new sources of chemicals from the oceans is now active in many parts of the world, and it is likely that activity will increase significantly in the coming years. This work aims to highlight the drivers, opportunities, challenges, and wider consequences of utilising these emerging classes of resource.

References

- Seven-reasons-ocean-important, sail-green: heoceanpreneur.com, Retrieved 15 June, 2019

- Env-container, environment, topics see-the-sea.org, Retrieved 14 March, 2019

- Charette, Matthew; Smith, Walter H. F. (2010). "The volume of Earth's ocean". Oceanography. 23 (2): 112–114. Doi:10.5670/oceanog.2010.51. Retrieved 27 September 2012

- Ocean-resources, ocean-dumping, conservation: marinebio.org, Retrieved 16 July, 2019

- "Titan Likely To Have Huge Underground Ocean | Mind Blowing Science". Mindblowingscience.com. Retrieved 2012-11-08

- Future-ocean-resources, projects, topics-policy: royalsociety.org, Retrieved 17 August, 2019

Marine Organisms

2

- **Benthos**
- **Nekton**
- **Plankton**

Marine organisms include plants, animals and other organisms that live in the oceans. They also include communities of organisms like Benthos, Nekton, Plankton, etc. The chapter closely examines these types of marine organisms as well as their subtypes to provide an extensive understanding of the subject.

The smallest and largest animals on Earth live in the oceans. Why do you think the oceans can support large animals?

Marine animals breathe air or extract oxygen from the water. Some float on the surface and others dive into the ocean's depths. There are animals that eat other animals, and plants generate food from sunlight. A few bizarre creatures break down chemicals to make food! The following section divides ocean life into seven basic groups.

Plankton

Plankton are organisms that cannot swim but that float along with the current. The word "plankton" comes from the Greek for wanderer. Most plankton are microscopic, but some are visible to the naked eye.

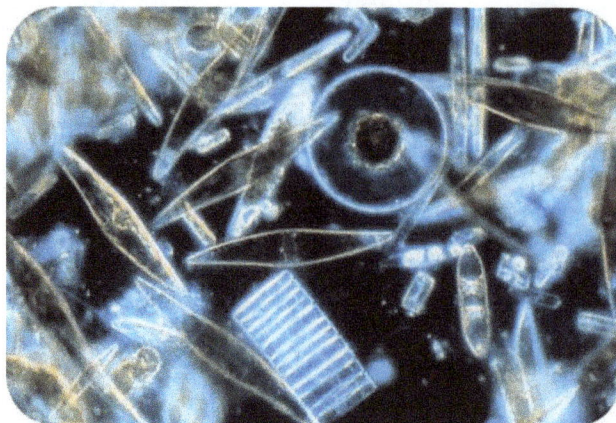

Microscopic diatoms are a type of phytoplankton.

Phytoplankton are tiny plants that make food by photosynthesis. Because they need sunlight, phytoplankton live in the photic zone. Phytoplankton are responsible for about half of the total primary productivity (food energy) on Earth. Like other plants, phytoplankton release oxygen as a waste product

Zooplankton, or animal plankton, eat phytoplankton as their source of food. Some zooplankton live as plankton all their lives and others are juvenile forms of animals that will attach to the bottom as adults. Some small invertebrates live as zooplankton.

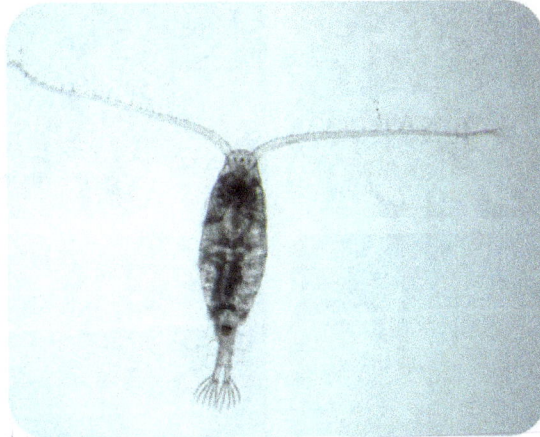

Copepods are abundant and so are an important food source for larger animals.

Plants and Algae

The few true plants found in the oceans include salt marsh grasses and mangrove trees. Although they are not true plants, large algae, which are called seaweed, also use photosynthesis to make food. Plants and seaweeds are found in the neritic zone, where the light they need penetrates so that they can photosynthesize.

Kelp grows in forests in the neritic zone. Otters and other organisms depend on the kelp-forest ecosystem.

Marine Invertebrates

The variety and number of invertebrates, animals without a backbone, is truly remarkable. Marine invertebrates include sea slugs, sea anemones, starfish, octopuses, clams, sponges, sea worms, crabs, and lobsters. Most of these animals are found close to the shore, but they can be found throughout the ocean.

(a) Mussels; (b) Crown of thorns sea star; (c) Moon jelly; (d) A squid.

Jellies are otherworldly creatures that glow in the dark, without brains or bones, some more than 100 feet long. Along with many other ocean areas, they live just off California's coast.

Fish

Fish are vertebrates; they have a backbone. What are some of the features fish have that allows them to live in the oceans? All fish have most or all of these traits:

- Fins with which to move and steer.

- Scales for protection.

- Gills for extracting oxygen from the water.

- A swim bladder that lets them rise and sink to different depths.

- Ectothermy (cold-bloodedness), so that their bodies are the same temperature as the surrounding water.

- Bioluminescence, or light created from a chemical reaction that can attract prey or mates in the dark ocean.

Included among the fish are sardines, salmon, and eels, as well as the sharks and rays (which lack swim bladders).

The Great White Shark is a fish that preys on other fish and marine mammals.

Reptiles

Only a few types of reptiles live in the oceans and they live in warm water. Why are reptiles so restricted in their ability to live in the sea? Sea turtles, sea snakes, saltwater crocodiles, and marine iguana that are found only at the Galapagos Islands sum up the marine reptile groups. Sea snakes bear live young in the ocean, but turtles, crocodiles, and marine iguanas all lay their eggs on land.

Seabirds

Many types of birds are adapted to living in the sea or on the shore. With their long legs for wading and long bills for digging in sand for food, shorebirds are well adapted for the intertidal zone. Many seabirds live on land but go to sea to fish, such as gulls, pelicans, and frigate birds. Some birds, like albatross, spend months at sea and only come on shore to raise chicks

(a) Shorebirds; (b) Seabirds; (c) Albatross.

Marine Mammals

What are the common traits of mammals? Mammals are endothermic (warm-blooded) vertebrates that give birth to live young, feed them with milk, and have hair, ears, and a jaw bone with teeth.

What traits might mammals have to be adapted to life in the ocean?

- For swimming: Streamlined bodies, slippery skin or hair, fins.
- For warmth: Fur, fat, high metabolic rate, small surface area to volume, specialized blood system.
- For salinity: Kidneys that excrete salt, impervious skin.

Benthos

Benthos is the assemblage of organisms inhabiting the seafloor. Benthic epifauna live upon the seafloor or upon bottom objects; the so-called infauna live within the sediments of the seafloor. By far the best-studied benthos are the macrobenthos, those forms larger than 1 mm (0.04 inch), which are dominated by polychaete worms, pelecypods, anthozoans, echinoderms, sponges, ascidians, and crustaceans. Meiobenthos, those organisms between 0.1 and 1 mm in size, include polychaetes, pelecypods, copepods, ostracodes, cumaceans, nematodes, turbellarians, and foraminiferans. The microbenthos, smaller than 0.1 mm, include bacteria, diatoms, ciliates, amoeba, and flagellates.

The variety and abundance of the benthos vary with latitude, depth, water temperature and salinity, locally determined conditions such as the nature of the substrate, and ecological circumstances such as predation and competition. The principal food sources for the benthos are plankton and organic debris from land. In shallow water, larger algae are important, and, where light reaches the bottom, benthic photosynthesizing diatoms are also a significant food source. Hard and sandy substrates are populated by suspension feeders such as sponges and pelecypods. Softer bottoms are dominated by deposit eaters, of which the polychaetes are the most important. Fishes, starfish, snails, cephalopods, and the larger crustaceans are important predators and scavengers.

Food Sources

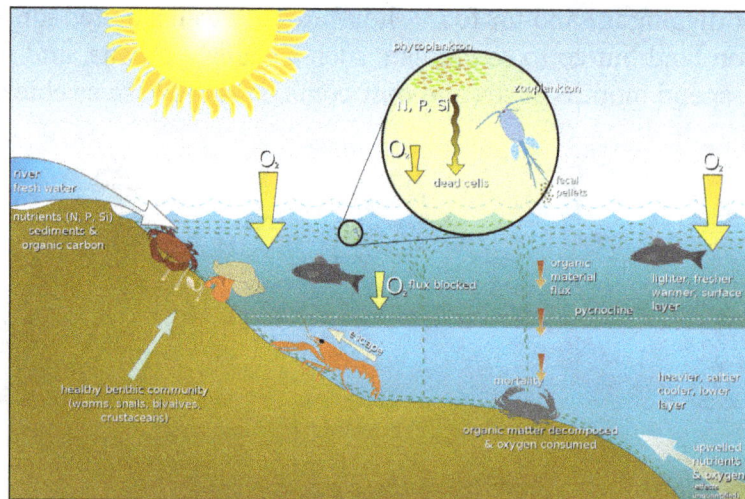

Effect of eutrophication on marine benthic life.

The main food sources for the benthos are algae and organic runoff from land. The depth of water, temperature and salinity, and type of local substrate all affect what benthos is present. In coastal waters and other places where light reaches the bottom, benthic photosynthesizing diatoms can proliferate. Filter feeders, such as sponges and bivalves, dominate hard, sandy bottoms. Deposit feeders, such as polychaetes, populate softer bottoms. Fish, such as dragonets, as well as sea stars, snails, cephalopods, and crustaceans are important predators and scavengers.

Benthic organisms, such as sea stars, oysters, clams, sea cucumbers, brittle stars and sea anemones, play an important role as a food source for fish, such as the California sheephead, and humans.

Macrobenthos

Macrobenthos comprises the larger, more visible, benthic organisms that are greater than 1 mm in size. Some examples are polychaete worms, bivalves, echinoderms, sea anemones, corals, sponges, sea squirts, turbellarians and larger crustaceans such as crabs, lobsters and cumaceans.

Echinoderms.

They are easily visible to the naked eye with the lower range of body size at 0.5 mm but usually larger than 3 mm. In the coastal water ecosystem, they include several species of organisms from different taxa including Porifera, Annelids, Coelenterates, Mollusks, Crustaceans, Arthropods etc.

A sea squirt being used as a substrate for a nudibranch's spiral egg.

Microphotograph of typical macrobenthic animals, (from top to bottom) including amphipods, a polychaete worm, a snail, and a chironomous midge larvae.

Meiobenthos

Meiobenthos comprises tiny benthic organisms that are less than 1 mm but greater than 0.1 mm in size. Some examples are nematodes, foraminiferans, water bears, gastrotriches and smaller crustaceans such as copepods and ostracodes.

Live foraminifera Ammonia tepida (Rotaliida).

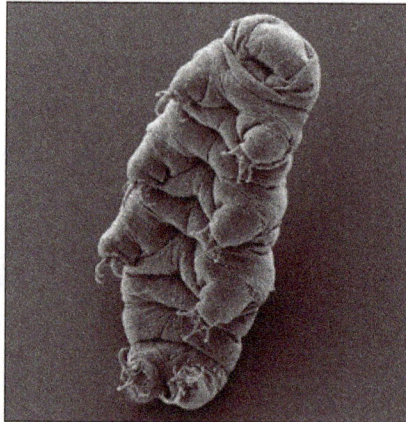
Water bear Hypsibius dujardini.

Microbenthos

Microbenthos comprises microscopic benthic organisms that are less than 0.1 mm in size. Some examples are bacteria, diatoms, ciliates, amoeba, flagellates.

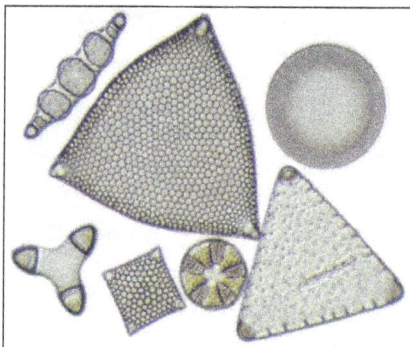
Marine diatoms.

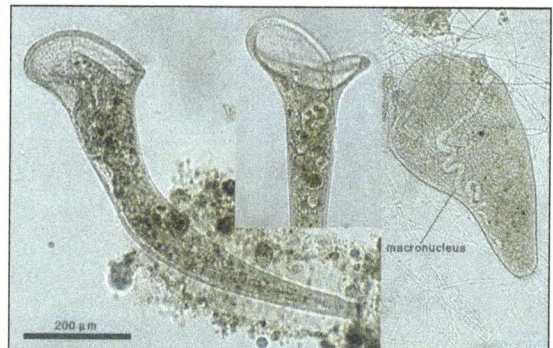
Ciliate stentor roeselii.

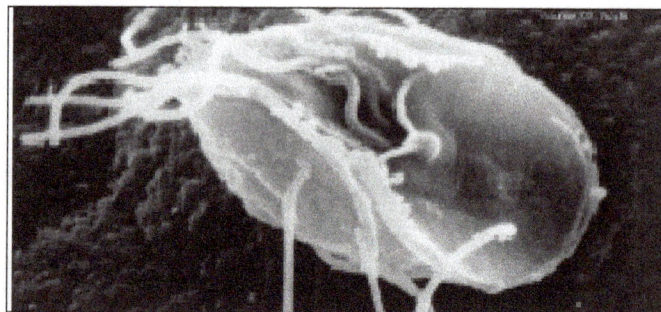
Flagellate.

Zoobenthos: Zoobenthos comprises the animals belonging to the benthos.

Phytobenthos: Phytobenthos comprises the plants belonging to the benthos, mainly benthic diatoms and macroalgae (seaweed).

Endobenthos: Endobenthos lives buried, or burrowing in the sediment, often in the oxygenated top layer, e.g., a sea pen or a sand dollar.

Epibenthos: Epibenthos lives on top of the sediments, e.g., like a sea cucumber or a sea snail crawling about. Unlike other epiphytes.

Hyperbenthos: Hyperbenthos lives just above the sediment, e.g., a rock cod.

Macrobenthos

Macrobenthos consists of the organisms that live at the bottom of a water column and are visible to the naked eye. In some classification schemes, these organisms are larger than 1 mm; in another, the smallest dimension must be at least 0.5 mm. They include polychaete worms, pelecypods, anthozoans, echinoderms, sponges, ascidians, crustaceans.

A visual examination of macroorganisms at the bottom of an aquatic ecosystem can be a good indicator of water quality.

Meiobenthos

Meiobenthos, also called meiofauna, are small benthic invertebrates that live in both marine and fresh water environments. The term *meiofauna* loosely defines a group of organisms by their size, larger than microfauna but smaller than macrofauna, rather than a taxonomic grouping. In practice, that is organisms that can pass through a 1 mm mesh but will be retained by a 45 μm mesh, but the exact dimensions will vary from researcher to researcher. Whether an organism will pass through a 1 mm mesh will also depend upon whether it is alive or dead at the time of sorting.

The term *meiobenthos* was first coined in 1942 by Mare, but organisms that would fit into the meiofauna category have been studied since the 18th century.

Collecting the Meiobenthos

Meiofauna are most commonly encountered in sedimentary environments in both marine and fresh water environments, from the littoral to the deep-sea. They can also be found on hard substrates living on algae, the phytal environment, and sessile animals (barnacles, mussel beds, etc.).

Sampling Methodologies

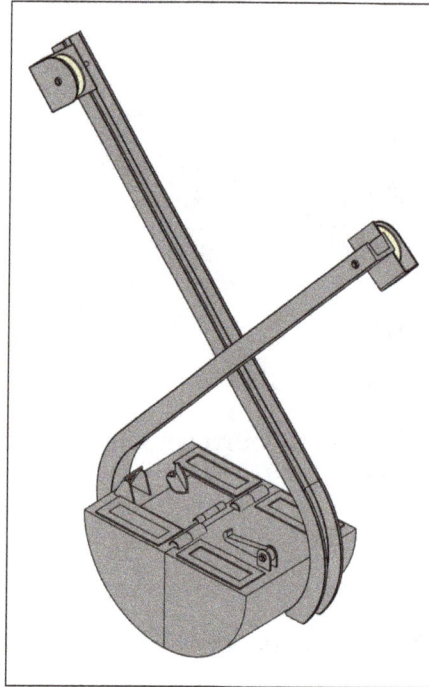

Van Veen grab.

Sampling the meiobenthos is clearly dependent upon the environment and whether quantitative or qualititative samples are required. In the sedimentary environment the methodology used also depends on the physical morphology of the sediment. For qualititative sampling within the littoral zone, for both coarse and fine sediment, a bucket and spade will work. In the sub-littoral and deep water some form of grab (like the Van Veen Grab Sampler) is required, although a fine mesh (about 0.25 mm or less) would work also.

For the quantitative sampling of sedimentary environments at all depths a wide variety of samplers have been devised. The simplest is a plastic syringe with the end cut off to form a piston corer which can be deployed in the littoral zone, or in the sub-littoral using SCUBA gear. Generally the deeper the water the more complicated the sampling process becomes. For sampling the meiofauna on hard substrates, phytal and epizooic environments, the only practical methodology is to cut or scrape off a known area of substrate and place it in a plastic bag.

Extraction Methodologies

There are a wide variety of methods for extracting meiofauna from the samples of their habitat depending upon whether live or fixed specimens are required. For extracting live meiofauna one has to contend with the large number of species that cling or attach themselves to the substrate when disturbed. In order to get the meiofauna to release their grip there are three methodologies available.

The first, and simplest, is osmotic shock, this is achieved by submerging the sample in fresh water (clearly this will only work for marine samples) for a few seconds. This will cause the organisms to release after which they can then be shaken free from the substrate and filtered out through a 45 µm mesh and immediately returned to fresh filtered seawater. Many organisms will come through this process unharmed as long as the osmotic shock does not last too long.

The second methodology is the use of an anaesthetic. The preferred solution for meiobenthologists is isotonic magnesium chloride (7.5g $MgCl_2 \cdot 6H_2O$ in 100 ml of distilled water). The sample is immersed in the isotonic solution and left for a period of 15 min, after which the meiofauna are shaken free of the substrate and again filtered out through a 45 µm mesh and immediately returned to fresh filtered seawater.

The third methodology is Uhlig's seawater ice technique. This relies on the organisms moving ahead of a front of ice cold seawater moving down through the sample ultimately forcing them out of the sediment. It is most effective on samples from temperate and tropical regions.

For major studies where large numbers of samples are collected concurrently, samples are normally fixed using 10% formalin solution and the meiofauna extracted at a later date. There are two main extraction methodologies. The first, decantation, works best with coarse sediments. Samples are shaken in an excess of water, the sediment is briefly allowed to settle and the meiofauna filtered off. The second methodology, the flotation technique, works best with finer sediments where the mass of the sediment particles is close to that of the meiofauna. The best solution to use is the colloidal silica, Ludox. The sample is stirred into the Ludox solution and left to settle for 40 min, after which the meiofauna are filtered out. In fine sediments, extraction efficiency is improved with centrifugation. With both methodologies repeated extractions should be made (at least three) with each sample to ensure that at least 95% of the fauna is extracted.

Nekton

Nekton is the assemblage of pelagic animals that swim freely, independent of water motion or wind. Only three phyla are represented by adult forms. Chordate nekton include numerous species of bony fishes, the cartilaginous fishes such as the sharks, several species of reptiles (turtles, snakes, and saltwater crocodiles), and mammals such as the whales, porpoises, and seals. Molluscan nekton include the squids and octopods. The only arthropod nekton are decapods, including shrimps, crabs, and lobsters.

Herbivorous nekton are not very common, although a few nearshore and shallow-water species subsist by grazing on plants. Of the nektonic feeding types, zooplankton feeders are the most abundant and include, in addition to many bony fishes, such as the sardines and mackerel, some of the largest nekton, the baleen whales. The molluscans, sharks, and many of the larger bony fishes consume animals bigger than zooplankton. Other fishes and most of the crustaceans are scavengers.

Nektonic species are limited in their areal and vertical distributions by the barriers of temperature, salinity, nutrient supply, and type of sea bottom. The number of nektonic species and individuals decreases with increasing depth in the ocean.

As a guideline, nektonic organisms have a high Reynolds number (greater than 1000) and planktonic organisms a low one (less than 10). However, some organisms can begin life as plankton and transition to nekton later on in life, sometimes making distinction difficult when attempting to classify certain plankton-to-nekton species as one or the other. For this reason, some biologists choose not to use this term.

The term was first proposed and used by the German biologist Ernst Haeckel in 1891, he contrasted it with plankton, the aggregate of passively floating, drifting, or somewhat motile organisms present in a body of water, primarily tiny algae and bacteria, small eggs and larvae of marine organisms, and protozoa and other minute consumers. Today it is sometimes considered an obsolete term because it often does not allow for the meaningful quantifiable distinction between these two groups. Some biologists no longer use it.

Oceanic Nekton

Oceanic nekton comprises animals largely from three clades:

- Vertebrates form the largest contribution; these animals are supported by either bones or cartilage.

- Mollusks are animals such as squids and scallops.

- Crustaceans are animals such as lobsters and crabs.

- There are organisms whose initial life stage is identified as being planktonic but when they grow and increase in body size they become nektonic. A typical example is the medusa of the jellyfish.

Mollusca

Mollusca is the second-largest phylum of invertebrate animals. The members are known as molluscs or mollusks. Around 85,000 extant species of molluscs are recognized. The number of fossil species is estimated between 60,000 and 100,000 additional species. The proportion of undescribed species is very high. Many taxa remain poorly studied.

Molluscs are the largest marine phylum, comprising about 23% of all the named marine organisms. Numerous molluscs also live in freshwater and terrestrial habitats. They are highly diverse, not just in size and anatomical structure, but also in behaviour and habitat. The phylum is typically divided into 8 or 9 taxonomic classes, of which two are entirely extinct. Cephalopod molluscs, such as squid, cuttlefish, and octopuses, are among the most neurologically advanced of all invertebrates—and either the giant squid or the colossal squid is the largest known invertebrate species. The gastropods (snails and slugs) are by far the most numerous molluscs and account for 80% of the total classified species.

The three most universal features defining modern molluscs are a mantle with a significant cavity used for breathing and excretion, the presence of a radula (except for bivalves), and the structure of the nervous system. Other than these common elements, molluscs express great morphological diversity, so many textbooks base their descriptions on a "hypothetical ancestral mollusc". This has a single, "limpet-like" shell on top, which is made of proteins and chitin reinforced with calcium carbonate, and is secreted by a mantle covering the whole upper surface. The underside of the animal consists of a single muscular "foot". Although molluscs are coelomates, the coelom tends to be small. The main body cavity is a hemocoel through which blood circulates; as such, their circulatory systems are mainly open. The "generalized" mollusc's feeding system consists of a rasping "tongue", the radula, and a complex digestive system in which exuded mucus and microscopic, muscle-powered "hairs" called cilia play various important roles. The generalized mollusc has two paired nerve cords, or three in bivalves. The brain, in species that have one, encircles the esophagus. Most molluscs have eyes, and all have sensors to detect chemicals, vibrations, and touch. The simplest type of molluscan reproductive system relies on external fertilization, but more complex variations occur. All produce eggs, from which may emerge trochophore larvae, more complex veliger larvae, or miniature adults. The coelomic cavity is reduced. They have an open circulatory system and kidney-like organs for excretion.

Good evidence exists for the appearance of gastropods, cephalopods, and bivalves in the Cambrian period, 541 to 485.4 million years ago. However, the evolutionary history both of molluscs' emergence from the ancestral Lophotrochozoa and of their diversification into the well-known living and fossil forms are still subjects of vigorous debate among scientists.

Molluscs have been and still are an important food source for anatomically modern humans. A risk of food poisoning exists from toxins that can accumulate in certain molluscs under specific conditions, however, and because of this, many countries have regulations to reduce this risk. Molluscs have, for centuries, also been the source of important luxury goods, notably pearls, mother of pearl, Tyrian purple dye, and sea silk. Their shells have also been used as money in some preindustrial societies.

Mollusc species can also represent hazards or pests for human activities. The bite of the blue-ringed octopus is often fatal, and that of *Octopus apollyon* causes inflammation that can last over a month. Stings from a few species of large tropical cone shells can also kill, but their sophisticated, though easily produced, venoms have become important tools in neurological research. Schistosomiasis (also known as bilharzia, bilharziosis, or snail fever) is transmitted to humans by water snail hosts, and affects about 200 million people. Snails and slugs can also be serious agricultural pests, and accidental or deliberate introduction of some snail species into new environments has seriously damaged some ecosystems.

The most universal features of the body structure of molluscs are a mantle with a significant cavity used for breathing and excretion, and the organization of the nervous system. Many have a calcareous shell.

Molluscs have developed such a varied range of body structures, finding synapomorphies (defining characteristics) to apply to all modern groups is difficult. The most general characteristic of molluscs is they are unsegmented and bilaterally symmetrical. The following are present in all modern molluscs:

- The dorsal part of the body wall is a mantle (or pallium) which secretes calcareous spicules, plates or shells. It overlaps the body with enough spare room to form a mantle cavity.

- The anus and genitals open into the mantle cavity.

- There are two pairs of main nerve cords.

Other characteristics that commonly appear in textbooks have significant exceptions:

Supposed universal Molluscan characteristic	Whether characteristic is found in these classes of Molluscs						
	Aplacophora	Polyplacophora	Monoplacophora	Gastropoda	Cephalopoda	Bivalvia	Scaphopoda
Radula, a rasping "tongue" with chitinous teeth	Absent in 20% of Neomeniomorpha	Yes	Yes	Yes	Yes	No	Internal, cannot extend beyond body
Broad, muscular foot	Reduced or absent	Yes	Yes	Yes	Modified into arms	Yes	Small, only at "front" end
Dorsal concentration of internal organs (visceral mass)	Not obvious	Yes	Yes	Yes	Yes	Yes	Yes
Large digestive ceca	No ceca in some Aplacophora	Yes	Yes	Yes	Yes	Yes	No
Large complex metanephridia ("kidneys")	None	Yes	Yes	Yes	Yes	Yes	Small, simple
One or more valves/ shells	Primitive forms, yes; modern forms, no	Yes	Yes	Snails, yes; slugs, mostly yes (internal vestigial)	Octopuses, no; cuttlefish, nautilus, squid, yes	Yes	Yes
Odontophore	Yes	Yes	Yes	Yes	Yes	No	Yes

Diversity

Diversity and variability of shells of molluscs on display.

About 80% of all known mollusc species are gastropods (snails and slugs), including this cowry (a sea snail).

Estimates of accepted described living species of molluscs vary from 50,000 to a maximum of 120,000 species. In 1969 David Nicol estimated the probable total number of living mollusc species at 107,000 of which were about 12,000 fresh-water gastropods and 35,000 terrestrial. The Bivalvia would comprise about 14% of the total and the other five classes less than 2% of the living molluscs. In 2009, Chapman estimated the number of described living species at 85,000. Haszprunar in 2001 estimated about 93,000 named species, which include 23% of all named marine organisms. Molluscs are second only to arthropods in numbers of living animal species—far behind the arthropods' 1,113,000 but well ahead of chordates' 52,000. About 200,000 living species in total are estimated, and 70,000 fossil species, although the total number of mollusc species ever to have existed, whether or not preserved, must be many times greater than the number alive today.

Molluscs have more varied forms than any other animal phylum. They include snails, slugs and other gastropods; clams and other bivalves; squids and other cephalopods; and other lesser-known but similarly distinctive subgroups. The majority of species still live in the oceans, from the seashores to the abyssal zone, but some form a significant part of the freshwater fauna and the terrestrial ecosystems. Molluscs are extremely diverse in tropical and temperate regions, but can be found at all latitudes. About 80% of all known mollusc species are gastropods. Cephalopoda such as squid, cuttlefish, and octopuses are among the neurologically most advanced of all invertebrates. The giant squid, which until recently had not been observed alive in its adult form, is one of the largest invertebrates, but a recently caught specimen of the colossal squid, 10 m (33 ft) long and weighing 500 kg (1,100 lb), may have overtaken it.

Freshwater and terrestrial molluscs appear exceptionally vulnerable to extinction. Estimates of the numbers of nonmarine molluscs vary widely, partly because many regions have not been thoroughly surveyed. There is also a shortage of specialists who can identify all the animals in any one area to species. However, in 2004 the IUCN Red List of Threatened Species included nearly 2,000 endangered nonmarine molluscs. For comparison, the great majority of mollusc species are marine, but only 41 of these appeared on the 2004 Red List. About 42% of recorded extinctions since the year 1500 are of molluscs, consisting almost entirely of nonmarine species.

Hypothetical Ancestral Mollusc

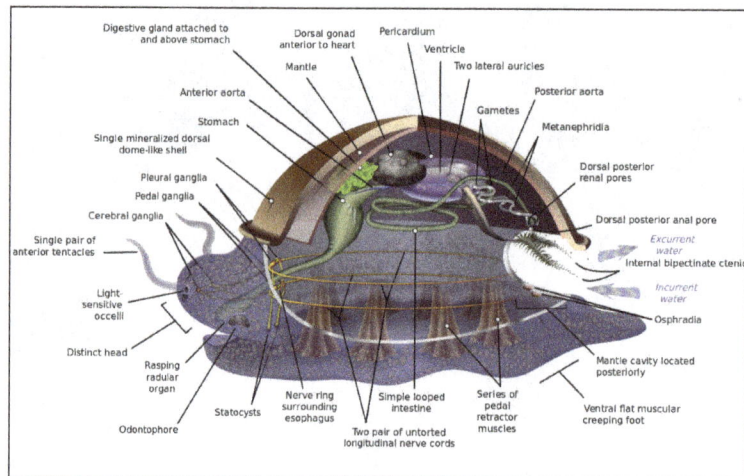

Anatomical diagram of a hypothetical ancestral mollusc.

Because of the great range of anatomical diversity among molluscs, many textbooks start the subject of molluscan anatomy by describing what is called an *archi-mollusc, hypothetical generalized mollusc,* or *hypothetical ancestral mollusc* (*HAM*) to illustrate the most common features found within the phylum. The depiction is visually rather similar to modern monoplacophorans.

The generalized mollusc is bilaterally symmetrical and has a single, "limpet-like" shell on top. The shell is secreted by a mantle covering the upper surface. The underside consists of a single muscular "foot". The visceral mass, or visceropallium, is the soft, nonmuscular metabolic region of the mollusc. It contains the body organs.

Mantle and Mantle Cavity

The mantle cavity, a fold in the mantle, encloses a significant amount of space. It is lined with epidermis, and is exposed, according to habitat, to sea, fresh water or air. The cavity was at the rear in the earliest molluscs, but its position now varies from group to group. The anus, a pair of osphradia (chemical sensors) in the incoming "lane", the hindmost pair of gills and the exit openings of the nephridia ("kidneys") and gonads (reproductive organs) are in the mantle cavity. The whole soft body of bivalves lies within an enlarged mantle cavity.

Shell

The mantle edge secretes a shell (secondarily absent in a number of taxonomic groups, such as the nudibranchs) that consists of mainly chitin and conchiolin (a protein hardened with calcium carbonate), except the outermost layer, which in almost all cases is all conchiolin. Molluscs never use phosphate to construct their hard parts, with the questionable exception of *Cobcrephora*. While most mollusc shells are composed mainly of aragonite, those gastropods that lay eggs with a hard shell use calcite (sometimes with traces of aragonite) to construct the eggshells.

The shell consists of three layers: the outer layer (the periostracum) made of organic matter, a middle layer made of columnar calcite, and an inner layer consisting of laminated calcite, often nacreous.

In some forms the shell contains openings. In abalones there are holes in the shell used for respiration and the release of egg and sperm, in the nautilus a string of tissue called the siphuncle goes through all the chambers, and the eight plates that make up the shell of chitons are penetrated with living tissue with nerves and sensory structures.

Foot

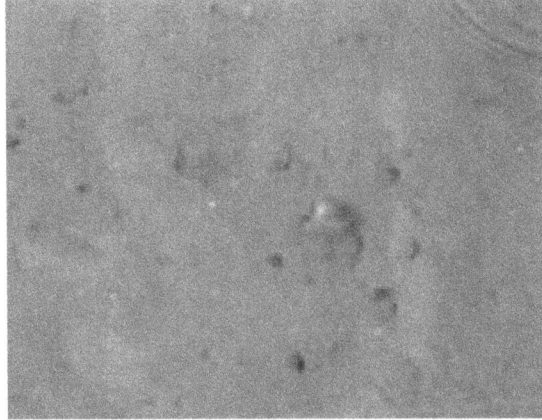

A 50-second snails (most likely *Natica chemnitzi* and *Cerithium stercusmuscaram*)
feeding on the sea floor in the Gulf of California, Puerto Peñasco, Mexico.

The underside consists of a muscular foot, which has adapted to different purposes in different classes. The foot carries a pair of statocysts, which act as balance sensors. In gastropods, it secretes mucus as a lubricant to aid movement. In forms having only a top shell, such as limpets, the foot acts as a sucker attaching the animal to a hard surface, and the vertical muscles clamp the shell down over it; in other molluscs, the vertical muscles pull the foot and other exposed soft parts into the shell. In bivalves, the foot is adapted for burrowing into the sediment; in cephalopods it is used for jet propulsion, and the tentacles and arms are derived from the foot.

Circulatory System

Most molluscs circulatory systems are mainly open. Although molluscs are coelomates, their coeloms are reduced to fairly small spaces enclosing the heart and gonads. The main body cavity is a hemocoel through which blood and coelomic fluid circulate and which encloses most of the other internal organs. These hemocoelic spaces act as an efficient hydrostatic skeleton. The blood of these molluscs contains the respiratory pigment hemocyanin as an oxygen-carrier. The heart consists of one or more pairs of atria (auricles), which receive oxygenated blood from the gills and pump it to the ventricle, which pumps it into the aorta (main artery), which is fairly short and opens into the hemocoel. The atria of the heart also function as part of the excretory system by filtering waste products out of the blood and dumping it into the coelom as urine. A pair of nephridia ("little kidneys") to the rear of and connected to the coelom extracts any re-usable materials from the urine and dumps additional waste products into it, and then ejects it via tubes that discharge into the mantle cavity.

Exceptions to the above are the molluscs *Planorbidae* or ram's horn snails, which are air-breathing snails that use iron-based hemoglobin instead of the copper-based hemocyanin to carry oxygen through their blood.

Respiration

Most molluscs have only one pair of gills, or even only a singular gill. Generally, the gills are rather like feathers in shape, although some species have gills with filaments on only one side. They divide the mantle cavity so water enters near the bottom and exits near the top. Their filaments have three kinds of cilia, one of which drives the water current through the mantle cavity, while the other two help to keep the gills clean. If the osphradia detect noxious chemicals or possibly sediment entering the mantle cavity, the gills' cilia may stop beating until the unwelcome intrusions have ceased. Each gill has an incoming blood vessel connected to the hemocoel and an outgoing one to the heart.

Eating, Digestion and Excretion

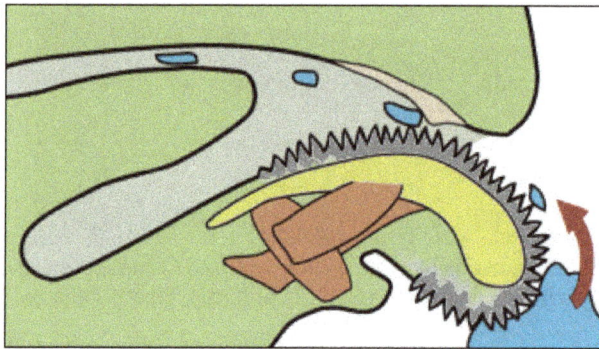

Snail radula at work= Food = Radula= Muscles= Odontophore "belt".

Members of the mollusc family use intracellular digestion to function. Most molluscs have muscular mouths with radulae, "tongues", bearing many rows of chitinous teeth, which are replaced from the rear as they wear out. The radula primarily functions to scrape bacteria and algae off rocks, and is associated with the odontophore, a cartilaginous supporting organ. The radula is unique to the molluscs and has no equivalent in any other animal.

Molluscs' mouths also contain glands that secrete slimy mucus, to which the food sticks. Beating cilia (tiny "hairs") drive the mucus towards the stomach, so the mucus forms a long string called a "food string".

At the tapered rear end of the stomach and projecting slightly into the hindgut is the prostyle, a backward-pointing cone of feces and mucus, which is rotated by further cilia so it acts as a bobbin, winding the mucus string onto itself. Before the mucus string reaches the prostyle, the acidity of the stomach makes the mucus less sticky and frees particles from it.

The particles are sorted by yet another group of cilia, which send the smaller particles, mainly minerals, to the prostyle so eventually they are excreted, while the larger ones, mainly food, are sent to the stomach's cecum (a pouch with no other exit) to be digested. The sorting process is by no means perfect.

Periodically, circular muscles at the hindgut's entrance pinch off and excrete a piece of the prostyle, preventing the prostyle from growing too large. The anus, in the part of the mantle cavity, is swept by the outgoing "lane" of the current created by the gills. Carnivorous molluscs usually have simpler digestive systems.

As the head has largely disappeared in bivalves, the mouth has been equipped with labial palps (two on each side of the mouth) to collect the detritus from its mucus.

Nervous System

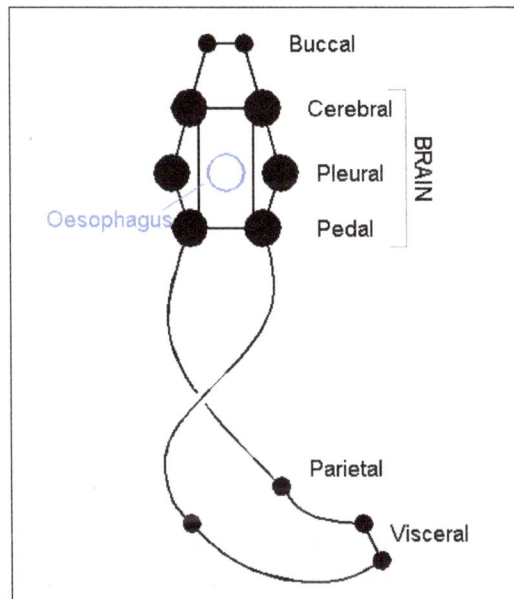

Simplified diagram of the mollusc nervous system.

The cephalic molluscs have two pairs of main nerve cords organized around a number of paired ganglia, the visceral cords serving the internal organs and the pedal ones serving the foot. Most pairs of corresponding ganglia on both sides of the body are linked by commissures (relatively large bundles of nerves). The ganglia above the gut are the cerebral, the pleural, and the visceral, which are located above the esophagus (gullet). The pedal ganglia, which control the foot, are below the esophagus and their commissure and connectives to the cerebral and pleural ganglia surround the esophagus in a circumesophageal nerve ring or *nerve collar*.

The acephalic molluscs (i.e., bivalves) also have this ring but it is less obvious and less important. The bivalves have only three pairs of ganglia— cerebral, pedal, and visceral— with the visceral as the largest and most important of the three functioning as the principal center of "thinking". Some such as the scallops have eyes around the edges of their shells which connect to a pair of looped nerves and which provide the ability to distinguish between light and shadow.

Reproduction

The simplest molluscan reproductive system relies on external fertilization, but with more complex variations. All produce eggs, from which may emerge trochophore larvae, more complex veliger larvae, or miniature adults. Two gonads sit next to the coelom, a small cavity that surrounds the heart, into which they shed ova or sperm. The nephridia extract the gametes from the coelom and emit them into the mantle cavity. Molluscs that use such a system remain of one sex all their lives and rely on external fertilization. Some molluscs use internal fertilization and are hermaphrodites, functioning as both sexes; both of these methods require more complex reproductive systems.

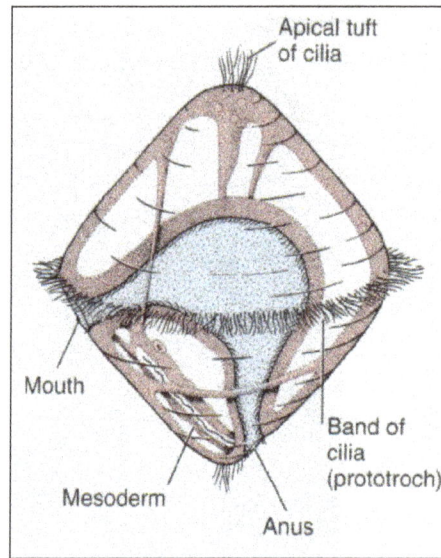

Trochophore larva.

The most basic molluscan larva is a trochophore, which is planktonic and feeds on floating food particles by using the two bands of cilia around its "equator" to sweep food into the mouth, which uses more cilia to drive them into the stomach, which uses further cilia to expel undigested remains through the anus. New tissue grows in the bands of mesoderm in the interior, so the apical tuft and anus are pushed further apart as the animal grows. The trochophore stage is often succeeded by a veliger stage in which the prototroch, the "equatorial" band of cilia nearest the apical tuft, develops into the velum ("veil"), a pair of cilia-bearing lobes with which the larva swims. Eventually, the larva sinks to the seafloor and metamorphoses into the adult form. While metamorphosis is the usual state in molluscs, the cephalopods differ in exhibiting direct development: the hatchling is a 'miniaturized' form of the adult.

Ecology

Feeding

Most molluscs are herbivorous, grazing on algae or filter feeders. For those grazing, two feeding strategies are predominant. Some feed on microscopic, filamentous algae, often using their radula as a 'rake' to comb up filaments from the sea floor. Others feed on macroscopic 'plants' such as kelp, rasping the plant surface with its radula. To employ this strategy, the plant has to be large enough for the mollusc to 'sit' on, so smaller macroscopic plants are not as often eaten as their larger counterparts. Filter feeders are molluscs that feed by straining suspended matter and food particle from water, typically by passing the water over their gills. Most bivalves are filter feeders.

Cephalopods are primarily predatory, and the radula takes a secondary role to the jaws and tentacles in food acquisition. The monoplacophoran *Neopilina* uses its radula in the usual fashion, but its diet includes protists such as the xenophyophore *Stannophyllum*. Sacoglossan sea-slugs suck the sap from algae, using their one-row radula to pierce the cell walls, whereas dorid nudibranchs and some Vetigastropoda feed on sponges and others feed on hydroids. (An extensive list of molluscs with unusual feeding habits is available in the appendix of GRAHAM, A.

Classification

Opinions vary about the number of classes of molluscs; for example, the table below shows seven living classes, and two extinct ones. Although they are unlikely to form a clade, some older works combine the Caudofoveata and Solenogasters into one class, the Aplacophora. Two of the commonly recognized "classes" are known only from fossils.

Class	Major organisms	Described living species	Distribution
Gastropoda	All the snails and slugs including abalone, limpets, conch, nudibranchs, sea hares, sea butterfly	70,000	Marine, freshwater, land
Bivalvia	Clams, oysters, scallops, geoducks, mussels	20,000	Marine, freshwater
Polyplacophora	Chitons	1,000	Rocky tidal zone and seabed
Cephalopoda	Squid, octopus, cuttlefish, nautilus, spirula	900	Marine
Scaphopoda	Tusk shells	500	Marine 6–7,000 metres (20–22,966 ft)
Aplacophora	Worm-like organisms	320	Seabed 200–3,000 metres (660–9,840 ft)
Monoplacophora	An ancient lineage of molluscs with cap-like shells	31	Seabed 1,800–7,000 metres (5,900–23,000 ft); one species 200 metres (660 ft)
Rostroconchia	Fossils; probable ancestors of bivalves	Extinct	Marine
Helcionelloida	Fossils; snail-like organisms such as latouchella	Extinct	Marine

Classification into higher taxa for these groups has been and remains problematic. A phylogenetic study suggests the Polyplacophora form a clade with a monophyletic Aplacophora. Additionally, it suggests a sister taxon relationship exists between the Bivalvia and the Gastropoda. Tentaculita may also be in Mollusca.

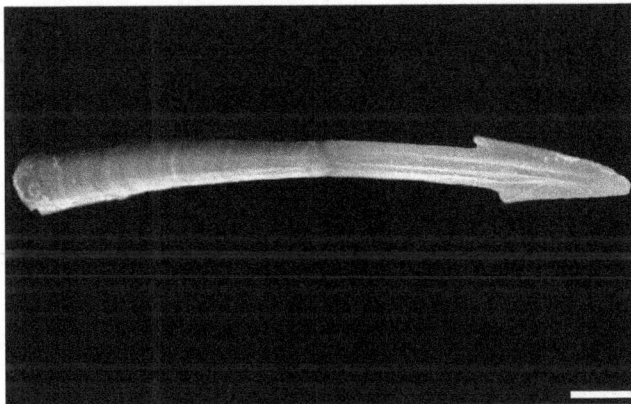

The use of love darts by the land snail *Monachoides vicinus* is a form of sexual selection.

Fossil Record

Good evidence exists for the appearance of gastropods (e.g. *Aldanella*), cephalopods (e.g. *Plectronoceras*, ?*Nectocaris*) and bivalves (*Pojetaia*, *Fordilla*) towards the middle of the Cambrian period, c. 500 million years ago, though arguably each of these may belong only to the stem lineage of their respective classes. However, the evolutionary history both of the emergence of molluscs from the ancestral group Lophotrochozoa, and of their diversification into the well-known living and fossil forms, is still vigorously debated.

Debate occurs about whether some Ediacaran and Early Cambrian fossils really are molluscs. *Kimberella*, from about 555 million years ago, has been described by some paleontologists as "mollusc-like", but others are unwilling to go further than "probable bilaterian", if that.

There is an even sharper debate about whether *Wiwaxia*, from about 505 million years ago, was a mollusc, and much of this centers on whether its feeding apparatus was a type of radula or more similar to that of some polychaete worms. Nicholas Butterfield, who opposes the idea that *Wiwaxia* was a mollusc, has written that earlier microfossils from 515 to 510 million years ago are fragments of a genuinely mollusc-like radula. This appears to contradict the concept that the ancestral molluscan radula was mineralized.

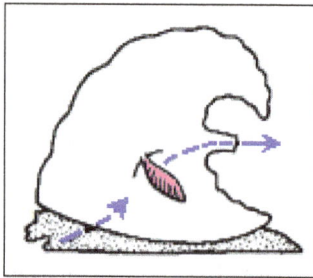

The tiny Helcionellid fossil Yochelcionella is thought to be an early mollusc

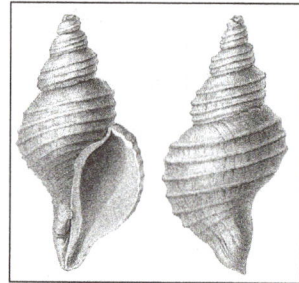

Spirally coiled shells appear in many gastropods.

However, the Helcionellids, which first appear over 540 million years ago in Early Cambrian rocks from Siberia and China, are thought to be early molluscs with rather snail-like shells. Shelled molluscs therefore predate the earliest trilobites. Although most helcionellid fossils are only a few millimeters long, specimens a few centimeters long have also been found, most with more limpet-like shapes. The tiny specimens have been suggested to be juveniles and the larger ones adults.

Some analyses of helcionellids concluded these were the earliest gastropods. However, other scientists are not convinced these Early Cambrian fossils show clear signs of the torsion that identifies modern gastropods twists the internal organs so the anus lies above the head.

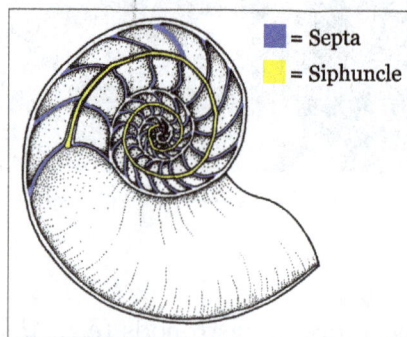

■ = Septa
■ = Siphuncle

Septa and siphuncle in nautiloid shell.

Volborthella, some fossils of which predate 530 million years ago, was long thought to be a cephalopod, but discoveries of more detailed fossils showed its shell was not secreted, but built from grains of the mineral silicon dioxide (silica), and it was not divided into a series of compartments by septa as those of fossil shelled cephalopods and the living *Nautilus* are. *Volborthella*'s classification is uncertain. The Late Cambrian fossil *Plectronoceras* is now thought to be the earliest clearly cephalopod fossil, as its shell had septa and a siphuncle, a strand of tissue that *Nautilus* uses to remove water from compartments it has vacated as it grows, and which is also visible in fossil ammonite shells. However, *Plectronoceras* and other early cephalopods crept along the seafloor instead of swimming, as their shells contained a "ballast" of stony deposits on what is thought to be the underside, and had stripes and blotches on what is thought to be the upper surface. All cephalopods with external shells except the nautiloids became extinct by the end of the Cretaceous period 65 million years ago. However, the shell-less Coleoidea (squid, octopus, cuttlefish) are abundant today.

The Early Cambrian fossils *Fordilla* and *Pojetaia* are regarded as bivalves. "Modern-looking" bivalves appeared in the Ordovician period, 488 to 443 million years ago. One bivalve group, the rudists, became major reef-builders in the Cretaceous, but became extinct in the Cretaceous–Paleogene extinction event. Even so, bivalves remain abundant and diverse.

The Hyolitha are a class of extinct animals with a shell and operculum that may be molluscs. Authors who suggest they deserve their own phylum do not comment on the position of this phylum in the tree of life.

Phylogeny

The phylogeny (evolutionary "family tree") of molluscs is a controversial subject. In addition to the debates about whether *Kimberella* and any of the "halwaxiids" were molluscs or closely related to molluscs, debates arise about the relationships between the classes of living molluscs. In fact, some groups traditionally classified as molluscs may have to be redefined as distinct but related.

Molluscs are generally regarded members of the Lophotrochozoa, a group defined by having trochophore larvae and, in the case of living Lophophorata, a feeding structure called a lophophore. The other members of the Lophotrochozoa are the annelid worms and seven marine phyla. The diagram on the right summarizes a phylogeny presented in 2007.

Because the relationships between the members of the family tree are uncertain, it is difficult to identify the features inherited from the last common ancestor of all molluscs. For example, it is uncertain whether the ancestral mollusc was metameric (composed of repeating units)—if it was, that would suggest an origin from an annelid-like worm. Scientists disagree about this: Giribet and colleagues concluded, in 2006, the repetition of gills and of the foot's retractor muscles were later developments, while in 2007, Sigwart concluded the ancestral mollusc was metameric, and it had a foot used for creeping and a "shell" that was mineralized. In one particular branch of the family tree, the shell of conchiferans is thought to have evolved from the spicules (small spines) of aplacophorans; but this is difficult to reconcile with the embryological origins of spicules.

The molluscan shell appears to have originated from a mucus coating, which eventually stiffened into a cuticle. This would have been impermeable and thus forced the development of more sophisticated respiratory apparatus in the form of gills. Eventually, the cuticle would have become mineralized, using the same genetic machinery (engrailed) as most other bilaterian skeletons. The first mollusc shell almost certainly was reinforced with the mineral aragonite.

The evolutionary relationships within the molluscs are also debated, and the diagrams below show two widely supported reconstructions:

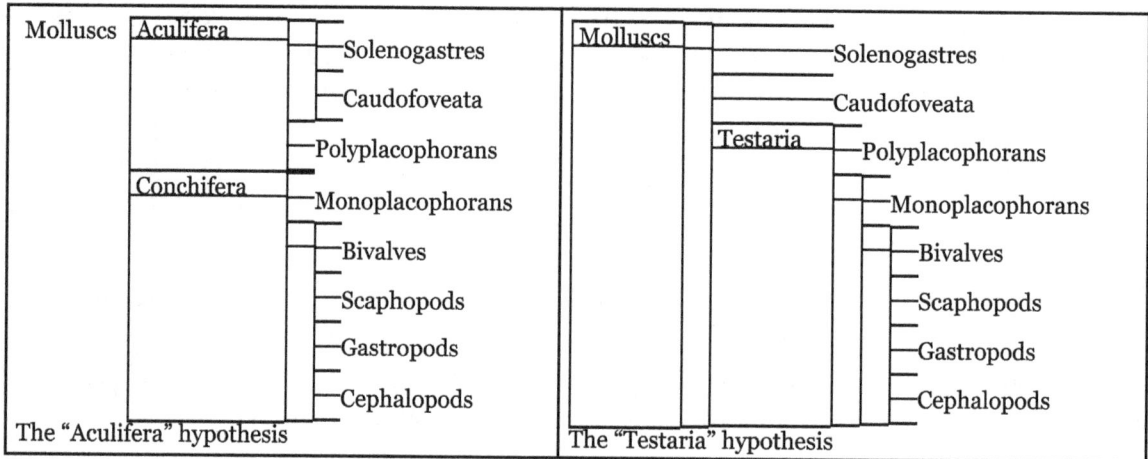

Morphological analyses tend to recover a conchiferan clade that receives less support from molecular analyses, although these results also lead to unexpected paraphylies, for instance scattering the bivalves throughout all other mollusc groups.

However, an analysis in 2009 using both morphological and molecular phylogenetics comparisons concluded the molluscs are not monophyletic; in particular, Scaphopoda and Bivalvia are both separate, monophyletic lineages unrelated to the remaining molluscan classes; the traditional phylum Mollusca is polyphyletic, and it can only be made monophyletic if scaphopods and bivalves are excluded. A 2010 analysis recovered the traditional conchiferan and aculiferan groups, and showed molluscs were monophyletic, demonstrating that available data for solenogastres was contaminated. Current molecular data are insufficient to constrain the molluscan phylogeny, and since the methods used to determine the confidence in clades are prone to overestimation, it is risky to place too much emphasis even on the areas of which different studies agree. Rather than eliminating unlikely relationships, the latest studies add new permutations of internal molluscan relationships, even bringing the conchiferan hypothesis into question.

Human Interaction

For millennia, molluscs have been a source of food for humans, as well as important luxury goods, notably pearls, mother of pearl, Tyrian purple dye, sea silk, and chemical compounds. Their shells have also been used as a form of currency in some preindustrial societies. A number of species of molluscs can bite or sting humans, and some have become agricultural pests.

Uses by Humans

Molluscs, especially bivalves such as clams and mussels, have been an important food source since at least the advent of anatomically modern humans, and this has often resulted in overfishing.

Other commonly eaten molluscs include octopuses and squids, whelks, oysters, and scallops. In 2005, China accounted for 80% of the global mollusc catch, netting almost 11,000,000 tonnes (11,000,000 long tons; 12,000,000 short tons). Within Europe, France remained the industry leader. Some countries regulate importation and handling of molluscs and other seafood, mainly to minimize the poison risk from toxins that can sometimes accumulate in the animals.

Saltwater pearl oyster farm in Seram, Indonesia.

Most molluscs with shells can produce pearls, but only the pearls of bivalves and some gastropods, whose shells are lined with nacre, are valuable. The best natural pearls are produced by marine pearl oysters, *Pinctada margaritifera* and *Pinctada mertensi*, which live in the tropical and subtropical waters of the Pacific Ocean. Natural pearls form when a small foreign object gets stuck between the mantle and shell.

The two methods of culturing pearls insert either "seeds" or beads into oysters. The "seed" method uses grains of ground shell from freshwater mussels, and overharvesting for this purpose has endangered several freshwater mussel species in the southeastern United States. The pearl industry is so important in some areas, significant sums of money are spent on monitoring the health of farmed molluscs.

Byzantine Emperor Justinian I clad in Tyrian purple and wearing numerous pearls.

Other luxury and high-status products were made from molluscs. Tyrian purple, made from the ink glands of murex shells, "fetched its weight in silver" in the fourth century BC, according to Theopompus. The discovery of large numbers of *Murex* shells on Crete suggests the Minoans may have pioneered the extraction of "imperial purple" during the Middle Minoan period in the 20th–18th centuries BC, centuries before the Tyrians. Sea silk is a fine, rare, and valuable fabric produced from the long silky threads (byssus) secreted by several bivalve molluscs, particularly *Pinna nobilis*, to attach themselves to the sea bed. Procopius, writing on the Persian wars *circa* 550 CE, "stated that the five hereditary satraps (governors) of Armenia who received their insignia from the Roman Emperor were given chlamys (or cloaks) made from *lana pinna*. Apparently, only the ruling classes were allowed to wear these chlamys."

Mollusc shells, including those of cowries, were used as a kind of money (shell money) in several preindustrial societies. However, these "currencies" generally differed in important ways from the standardized government-backed and -controlled money familiar to industrial societies. Some shell "currencies" were not used for commercial transactions, but mainly as social status displays at important occasions, such as weddings. When used for commercial transactions, they functioned as commodity money, as a tradable commodity whose value differed from place to place, often as a result of difficulties in transport, and which was vulnerable to incurable inflation if more efficient transport or "goldrush" behavior appeared.

Bioindicators

Bivalve molluscs are used as bioindicators to monitor the health of aquatic environments in both fresh water and the marine environments. Their population status or structure, physiology, behaviour or the level of contamination with elements or compounds can indicate the state of contamination status of the ecosystem. They are particularly useful since they are sessile so that they are representative of the environment where they are sampled or placed. Potamopyrgus antipodarum is used by some water treatment plants to test for estrogen-mimicking pollutants from industrial agriculture.

Harmful to Humans

Stings and Bites

The blue-ringed octopus's rings are a warning signal; this octopus is alarmed, and its bite can kill.

Some molluscs sting or bite, but deaths from mollusc venoms total less than 10% of those from jellyfish stings.

All octopuses are venomous, but only a few species pose a significant threat to humans. Blue-ringed octopuses in the genus *Hapalochlaena*, which live around Australia and New Guinea, bite humans only if severely provoked, but their venom kills 25% of human victims. Another tropical species, *Octopus apollyon*, causes severe inflammation that can last for over a month even if treated correctly, and the bite of *Octopus rubescens* can cause necrosis that lasts longer than one month if untreated, and headaches and weakness persisting for up to a week even if treated.

Live cone snails can be dangerous to shell collectors, but are useful to neurology researchers.

All species of cone snails are venomous and can sting painfully when handled, although many species are too small to pose much of a risk to humans, and only a few fatalities have been reliably reported. Their venom is a complex mixture of toxins, some fast-acting and others slower but deadlier. The effects of individual cone-shell toxins on victims' nervous systems are so precise as to be useful tools for research in neurology, and the small size of their molecules makes it easy to synthesize them.

Disease Vectors

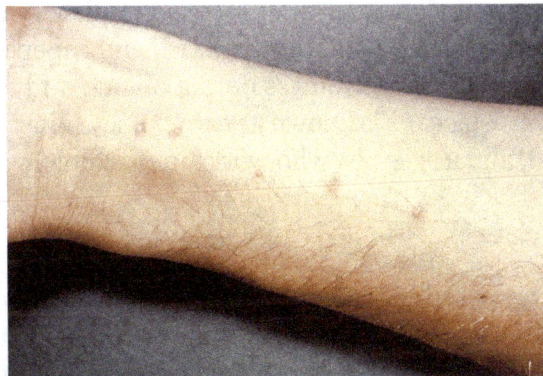

Skin vesicles created by the penetration of Schistosoma.

Schistosomiasis (also known as bilharzia, bilharziosis or snail fever), a disease caused by the fluke worm *Schistosoma*, is "second only to malaria as the most devastating parasitic disease in tropical countries. An estimated 200 million people in 74 countries are infected with the disease – 100 million in Africa alone." The parasite has 13 known species, two of which infect humans. The parasite itself is not a mollusc, but all the species have freshwater snails as intermediate hosts.

Pests

Some species of molluscs, particularly certain snails and slugs, can be serious crop pests, and when introduced into new environments, can unbalance local ecosystems. One such pest, the giant African snail *Achatina fulica*, has been introduced to many parts of Asia, as well as to many islands in the Indian Ocean and Pacific Ocean. In the 1990s, this species reached the West Indies. Attempts to control it by introducing the predatory snail *Euglandina rosea* proved disastrous, as the predator ignored *Achatina fulica* and went on to extirpate several native snail species, instead.

Crustacean

Crustaceans form a large, diverse arthropod taxon which includes such familiar animals as crabs, lobsters, crayfish, shrimps, prawns, krill, woodlice, and barnacles. The crustacean group is usually treated as a subphylum, and because of recent molecular studies it is now well accepted that the crustacean group is paraphyletic, and comprises all animals in the Pancrustacea clade other than hexapods. Some crustaceans are more closely related to insects and other hexapods than they are to certain other crustaceans.

The 67,000 described species range in size from *Stygotantulus stocki* at 0.1 mm (0.004 in), to the Japanese spider crab with a leg span of up to 3.8 m (12.5 ft) and a mass of 20 kg (44 lb). Like other arthropods, crustaceans have an exoskeleton, which they moult to grow. They are distinguished from other groups of arthropods, such as insects, myriapods and chelicerates, by the possession of biramous (two-parted) limbs, and by their larval forms, such as the nauplius stage of branchiopods and copepods.

Most crustaceans are free-living aquatic animals, but some are terrestrial (e.g. woodlice), some are parasitic (e.g. Rhizocephala, fish lice, tongue worms) and some are sessile (e.g. barnacles). The group has an extensive fossil record, reaching back to the Cambrian, and includes living fossils such as *Triops cancriformis*, which has existed apparently unchanged since the Triassic period. More than 7.9 million tons of crustaceans per year are produced by fishery or farming for human consumption, the majority of it being shrimp and prawns. Krill and copepods are not as widely fished, but may be the animals with the greatest biomass on the planet, and form a vital part of the food chain. The scientific study of crustaceans is known as carcinology (alternatively, *malacostracology*, *crustaceology* or *crustalogy*), and a scientist who works in carcinology is a carcinologist.

Structure

The body of a crustacean is composed of segments, which are grouped into three regions: the *cephalon* or head, the *pereon* or thorax, and the *pleon* or abdomen. The head and thorax may be fused together to form a cephalothorax, which may be covered by a single large carapace. The crustacean body is protected by the hard exoskeleton, which must be moulted for the animal to grow. The shell around each somite can be divided into a dorsal tergum, ventral sternum and a lateral pleuron. Various parts of the exoskeleton may be fused together.

A shed carapace of a lady crab, part of the hard exoskeleton.

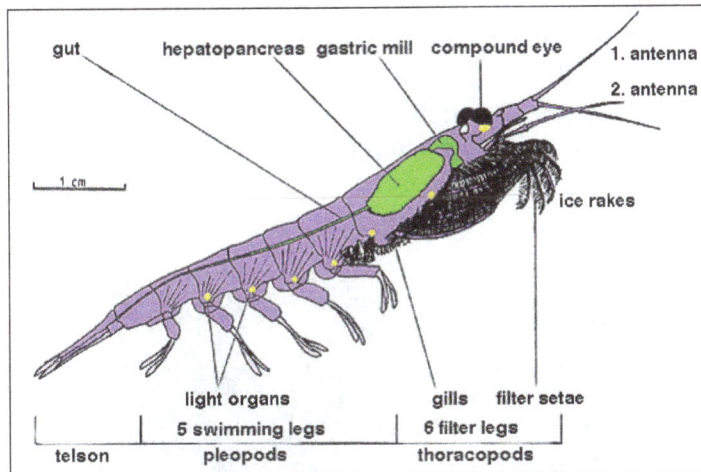

Body structure of a typical crustacean – krill.

Each somite, or body segment can bear a pair of appendages: on the segments of the head, these include two pairs of antennae, the mandibles and maxillae; the thoracic segments bear legs, which may be specialised as pereiopods (walking legs) and maxillipeds (feeding legs). The abdomen bears pleopods, and ends in a telson, which bears the anus, and is often flanked by uropods to form a tail fan. The number and variety of appendages in different crustaceans may be partly responsible for the group's success.

Crustacean appendages are typically biramous, meaning they are divided into two parts; this includes the second pair of antennae, but not the first, which is usually uniramous, the exception being in the Class Malacostraca where the antennules may be generally biramous or even triramous. It is unclear whether the biramous condition is a derived state which evolved in crustaceans, or whether the second branch of the limb has been lost in all other groups. Trilobites, for instance, also possessed biramous appendages.

The main body cavity is an open circulatory system, where blood is pumped into the haemocoel by a heart located near the dorsum. Malacostraca have haemocyanin as the oxygen-carrying pigment, while copepods, ostracods, barnacles and branchiopods have haemoglobins. The alimentary canal consists of a straight tube that often has a gizzard-like "gastric mill" for grinding food and a pair of digestive glands that absorb food; this structure goes in a spiral format. Structures that function as kidneys are located near the antennae. A brain exists in the form of ganglia close to the antennae, and a collection of major ganglia is found below the gut.

In many decapods, the first (and sometimes the second) pair of pleopods are specialised in the male for sperm transfer. Many terrestrial crustaceans (such as the Christmas Island red crab) mate seasonally and return to the sea to release the eggs. Others, such as woodlice, lay their eggs on land, albeit in damp conditions. In most decapods, the females retain the eggs until they hatch into free-swimming larvae.

Ecology

The majority of crustaceans are aquatic, living in either marine or freshwater environments, but a few groups have adapted to life on land, such as terrestrial crabs, terrestrial hermit crabs, and woodlice. Marine crustaceans are as ubiquitous in the oceans as insects are on land. The majority of crustaceans are also motile, moving about independently, although a few taxonomic units are parasitic and live attached to their hosts (including sea lice, fish lice, whale lice, tongue worms, and *Cymothoa exigua*, all of which may be referred to as "crustacean lice"), and adult barnacles live a sessile life – they are attached headfirst to the substrate and cannot move independently. Some branchiurans are able to withstand rapid changes of salinity and will also switch hosts from marine to non-marine species.

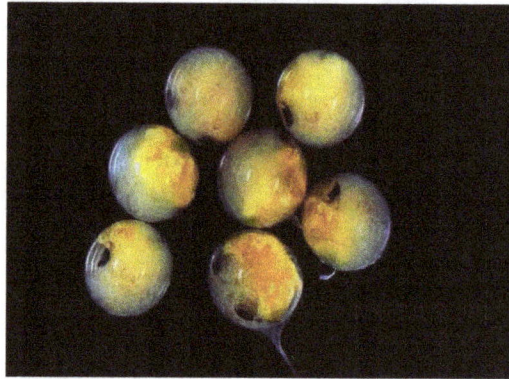

Eggs of *Potamon fluviatile*, a freshwater crab.

Zoea larva of the European lobster, Homarus gammarus.

Mating System

The majority of crustaceans have separate sexes, and reproduce sexually. A small number are hermaphrodites, including barnacles, remipedes, and Cephalocarida. Some may even change sex during the course of their life. Parthenogenesis is also widespread among crustaceans, where viable eggs are produced by a female without needing fertilisation by a male. This occurs in many branchiopods, some ostracods, some isopods, and certain "higher" crustaceans, such as the *Marmorkrebs* crayfish.

Eggs

In many groups of crustaceans, the fertilised eggs are simply released into the water column, while others have developed a number of mechanisms for holding on to the eggs until they are ready to hatch. Most decapods carry the eggs attached to the pleopods, while peracarids, notostracans, anostracans, and many isopods form a brood pouch from the carapace and thoracic limbs. Female Branchiura do not carry eggs in external ovisacs but attach them in rows to rocks and other objects. Most leptostracans and krill carry the eggs between their thoracic limbs; some copepods carry their eggs in special thin-walled sacs, while others have them attached together in long, tangled strings.

Larvae

Crustaceans exhibit a number of larval forms, of which the earliest and most characteristic is the nauplius. This has three pairs of appendages, all emerging from the young animal's head, and a single naupliar eye. In most groups, there are further larval stages, including the zoea. This name was given to it when naturalists believed it to be a separate species. It follows the nauplius stage and precedes the post-larva. Zoea larvae swim with their thoracic appendages, as opposed to nauplii, which use cephalic appendages, and megalopa, which use abdominal appendages for swimming. It often has spikes on its carapace, which may assist these small organisms in maintaining directional swimming. In many decapods, due to their accelerated development, the zoea is the first larval stage. In some cases, the zoea stage is followed by the mysis stage, and in others, by the megalopa stage, depending on the crustacean group involved.

Classification

The name "crustacean" dates from the earliest works to describe the animals, including those of Pierre Belon and Guillaume Rondelet, but the name was not used by some later authors, including Carl Linnaeus, who included crustaceans among the "Aptera" in his *Systema Naturae*. The earliest nomenclaturally valid work to use the name "Crustacea" was Morten Thrane Brünnich's *Zoologiæ Fundamenta* in 1772, although he also included chelicerates in the group.

The subphylum Crustacea comprises almost 67,000 described species, which is thought to be just 1/10 to 1/100 of the total number as the majority of species remain as yet undiscovered. Although most crustaceans are small, their morphology varies greatly and includes both the largest arthropod in the world – the Japanese spider crab with a leg span of 3.7 metres (12 ft) – and the smallest, the 100-micrometre-long (0.00004 in) *Stygotantulus stocki*. Despite their diversity of form, crustaceans are united by the special larval form known as the nauplius.

The exact relationships of the Crustacea to other taxa are not completely settled as of April 2012. Studies based on morphology led to the Pancrustacea hypothesis, in which Crustacea and Hexapoda (insects and allies) are sister groups. More recent studies using DNA sequences suggest that Crustacea is paraphyletic, with the hexapods nested within a larger Pancrustacea clade.

Although the classification of crustaceans has been quite variable, the system used by Martin and Davis largely supersedes earlier works. Mystacocarida and Branchiura, here treated as part of Maxillopoda, are sometimes treated as their own classes. Six classes are usually recognised:

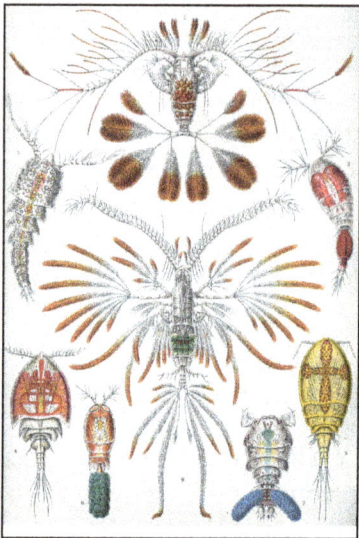

Copepods, from Ernst Haeckel's 1904 work Kunstformen der Natur.

Decapods, from Ernst Haeckel's 1904 work Kunstformen der Natur.

Fossil Record

Eryma mandelslohi, a fossil decapod from the Jurassic of Bissingen an der Teck, Germany.

Crustaceans have a rich and extensive fossil record, which begins with animals such as *Canadaspis* and *Perspicaris* from the Middle Cambrian age Burgess Shale. Most of the major groups of crustaceans appear in the fossil record before the end of the Cambrian, namely the Branchiopoda, Maxillopoda (including barnacles and tongue worms) and Malacostraca; there is some debate as to whether or not Cambrian animals assigned to Ostracoda are truly ostracods, which would otherwise start in the Ordovician. The only classes to appear later are the Cephalocarida, which have no fossil record, and the Remipedia, which were first described from the fossil *Tesnusocaris goldichi*, but do not appear until the Carboniferous. Most of the early crustaceans are rare, but fossil crustaceans become abundant from the Carboniferous period onwards.

Norway lobsters on sale at a Spanish market.

Within the Malacostraca, no fossils are known for krill, while both Hoplocarida and Phyllopoda contain important groups that are now extinct as well as extant members (Hoplocarida: mantis shrimp are extant, while Aeschronectida are extinct; Phyllopoda: Canadaspidida are extinct, while Leptostraca are extant). Cumacea and Isopoda are both known from the Carboniferous, as are the first true mantis shrimp. In the Decapoda, prawns and polychelids appear in the Triassic, and shrimp and crabs appear in the Jurassic; . The fossil burrow *Ophiomorpha* is attributed to ghost shrimps, whereas the fossil burrow *Camborygma* is attributed to crayfishes. The Permian–Triassic deposits of Nurra preserve the oldest (Permian: Roadian) fluvial burrows ascribed to ghost shrimps (Decapoda: Axiidea, Gebiidea) and crayfishes (Decapoda: Astacidea, Parastacidea), respectively.

However, the great radiation of crustaceans occurred in the Cretaceous, particularly in crabs, and may have been driven by the adaptive radiation of their main predators, bony fish. The first true lobsters also appear in the Cretaceous.

Consumption by Humans

Many crustaceans are consumed by humans, and nearly 10,700,000 tons were produced in 2007; the vast majority of this output is of decapod crustaceans: crabs, lobsters, shrimp, crawfish, and prawns. Over 60% by weight of all crustaceans caught for consumption are shrimp and prawns, and nearly 80% is produced in Asia, with China alone producing nearly half the world's total. Non-decapod crustaceans are not widely consumed, with only 118,000 tons of krill being caught, despite krill having one of the greatest biomasses on the planet.

Plankton

Plankton is marine and freshwater organisms that, because they are nonmotile or too small or weak to swim against the current, exist in a drifting state. The term plankton is a collective name for all such organisms—including certain algae, bacteria, protozoans, crustaceans, mollusks, and coelenterates, as well as representatives from almost every other phylum of animals. Plankton is distinguished from nekton, which is composed of strong-swimming animals, and from benthos, which includes sessile, creeping, and burrowing organisms on the seafloor. Large floating seaweeds (for example, Sargassum, which constitutes the Sargasso Sea) and various related multicellular algae are not considered plankton but pleuston. Pleuston are forms of life that live at the interface of air and water. Organisms resting or swimming on the surface film of the water are called neuston (e.g., the alga Ochromonas).

Plankton is the productive base of both marine and freshwater ecosystems, providing food for larger animals and indirectly for humans, whose fisheries depend upon plankton. As a human resource, plankton has only begun to be developed and exploited, in view of its high biological productivity and wide extent. It has been demonstrated on several occasions that large-scale cultures of algae are technically feasible. The unicellular green alga Chlorella has been used particularly in this connection. Through ample culture conditions, production is directed toward protein content greater than 50 percent. Although this protein has a suitable balance of essential amino acids, its low degree of digestibility prevents practical use. Phytoplankton may become increasingly important in space travel as a source for food and for gas exchange. The carbon dioxide released during respiration of spacecraft personnel would be transformed into organic substances by the algae, while the oxygen liberated during this process would support human respiration.

Plankton are the diverse collection of organisms that live in large bodies of water and are unable to swim against a current. The individual organisms constituting plankton are called plankters. They provide a crucial source of food to many large aquatic organisms, such as fish and whales.

These organisms include bacteria, archaea, algae, protozoa and drifting or floating animals that inhabit—for example—the pelagic zone of oceans, seas, or bodies of fresh water. Essentially, plankton are defined by their ecological niche rather than any phylogenetic or taxonomic classification.

Though many planktonic species are microscopic in size, *plankton* includes organisms over a wide range of sizes, including large organisms such as jellyfish. Technically the term does not include organisms on the surface of the water, which are called *pleuston*—or those that swim actively in the water, which are called *nekton*.

Trophic Groups

An amphipod (Hyperia macrocephala).

Plankton are primarily divided into broad functional (or trophic level) groups:

- Phytoplankton, are autotrophic prokaryotic or eukaryotic algae that live near the water surface where there is sufficient light to support photosynthesis. Among the more important groups are the diatoms, cyanobacteria, dinoflagellates and coccolithophores.

- Zooplankton, are small protozoans or metazoans (e.g. crustaceans and other animals) that feed on other plankton. Some of the eggs and larvae of larger nektonic animals, such as fish, crustaceans, and annelids, are included here.

- Bacterioplankton include bacteria and archaea, which play an important role in remineralising organic material down the water column (note that prokaryotic phytoplankton are also bacterioplankton).

- Mycoplankton, include fungi and fungus-like organisms, which, like bacterioplankton, are also significant in remineralisation and nutrient cycling.

- Mixotrophs. Plankton have traditionally been categorized as producer, consumer and recycler groups, but some plankton are able to benefit from more than just one trophic level. In this mixed trophic strategy — known as mixotrophy — organisms act as both producers and consumers, either at the same time or switching between modes of nutrition in response to ambient conditions. This makes it possible to use photosynthesis for growth when nutrients and light are abundant, but switching to eat phytoplankton, zooplankton or each other when growing conditions are poor. Mixotrophs are divided into two groups; constitutive mixotrophs, CMs, which are able to perform photosynthesis on their own, and non-constitutive mixotrophs, NCMs, which use phagocytosis to engulf phototrophic prey that are either kept alive inside the host cell which benefit from its photosynthesis, or they digest their prey except for the plastids which continues to perform photosynthesis (kleptoplasty).

Recognition of the importance of mixotrophy as an ecological strategy is increasing, as well as the wider role this may play in marine biogeochemistry. Studies have shown that mixotrophs are much more important for the marine ecology than previously assumed, and comprise more than half of all microscopic plankton.

Size Groups

Macroplankton: A Janthina janthina snail (with bubble float) cast up onto a beach in Maui.

Plankton are also often described in terms of size. Usually the following divisions are used:

Group	Size range (ESD)	Examples
Megaplankton	> 20 cm	Metazoans; e.g. jellyfish; ctenophores; salps and pyrosomes (pelagic Tunicata); Cephalopoda; Amphipoda.
Macroplankton	2→20 cm	Metazoans; e.g. Pteropods; Chaetognaths; Euphausiacea (krill); Medusae; ctenophores; salps, doliolids and pyrosomes (pelagic Tunicata); Cephalopoda; Janthinidae (one family of gastropods); Amphipoda.
Mesoplankton	0.2→20 mm	Metazoans; e.g. copepods; Medusae; Cladocera; Ostracoda; Chaetognaths; Pteropods; Tunicata.
Microplankton	20→200 μm	Large eukaryotic protists; most phytoplankton; Protozoa Foraminifera; tintinnids; other ciliates; Rotifera; juvenile metazoans - Crustacea (copepod nauplii).
Nanoplankton	2→20 μm	Small eukaryotic protists; Small Diatoms; Small Flagellates; Pyrrophyta; Chrysophyta; Chlorophyta; Xanthophyta.
Picoplankton	0.2→2 μm	Small eukaryotic protists; bacteria; Chrysophyta.
Femtoplankton	< 0.2 μm	Marine viruses.

However, some of these terms may be used with very different boundaries, especially on the larger end. The existence and importance of nano- and even smaller plankton was only discovered during the 1980s, but they are thought to make up the largest proportion of all plankton in number and diversity.

The microplankton and smaller groups are microorganisms and operate at low Reynolds numbers, where the viscosity of water is much more important than its mass or inertia.

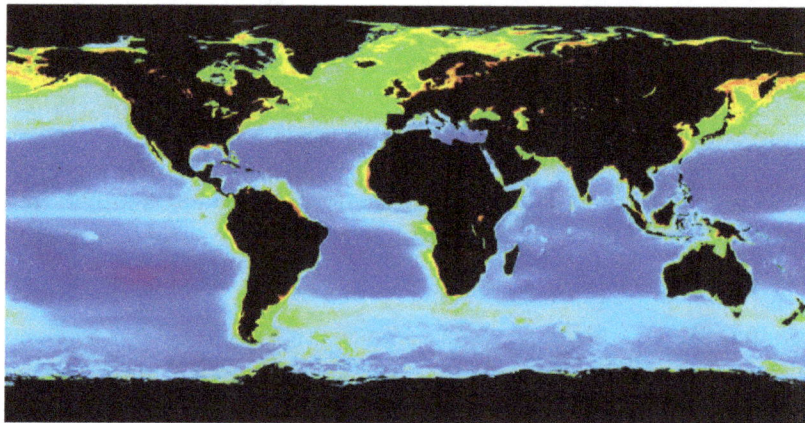

World concentrations of surface ocean chlorophyll as viewed by satellite during the northern spring, averaged from 1998 to 2004. Chlorophyll is a marker for the distribution and abundance of phytoplankton.

Plankton inhabit oceans, seas, lakes, ponds. Local abundance varies horizontally, vertically and seasonally. The primary cause of this variability is the availability of light. All plankton ecosystems are driven by the input of solar energy, confining primary production to surface waters, and to geographical regions and seasons having abundant light.

A secondary variable is nutrient availability. Although large areas of the tropical and sub-tropical oceans have abundant light, they experience relatively low primary production because they offer limited nutrients such as nitrate, phosphate and silicate. This results from large-scale ocean circulation and water column stratification. In such regions, primary production usually occurs at greater depth, although at a reduced level (because of reduced light).

Despite significant macronutrient concentrations, some ocean regions are unproductive (so-called HNLC regions). The micronutrient iron is deficient in these regions, and adding it can lead to the formation of phytoplankton blooms. Iron primarily reaches the ocean through the deposition of dust on the sea surface. Paradoxically, oceanic areas adjacent to unproductive, arid land thus typically have abundant phytoplankton (e.g., the eastern Atlantic Ocean, where trade winds bring dust from the Sahara Desert in north Africa).

While plankton are most abundant in surface waters, they live throughout the water column. At depths where no primary production occurs, zooplankton and bacterioplankton instead consume organic material sinking from more productive surface waters above. This flux of sinking material, so-called marine snow, can be especially high following the termination of spring blooms.

Ecological Significance

Food Chain

Aside from representing the bottom few levels of a food chain that supports commercially important fisheries, plankton ecosystems play a role in the biogeochemical cycles of many important chemical elements, including the ocean's carbon cycle.

Carbon Cycle

Primarily by grazing on phytoplankton, zooplankton provide carbon to the planktic foodweb, either respiring it to provide metabolic energy, or upon death as biomass or detritus. Organic material tends to be denser than seawater, so it sinks into open ocean ecosystems away from the coastlines, transporting carbon along with it. This process, called the *biological pump*, is one reason that oceans constitute the largest carbon sink on Earth. However, it has been shown to be influenced by increments of temperature.

It might be possible to increase the ocean's uptake of carbon dioxide (CO_2) generated through human activities by increasing plankton production through *seeding*, primarily with the micronutrient iron. However, this technique may not be practical at a large scale. Ocean oxygen depletion and resultant methane production (caused by the excess production remineralising at depth) is one potential drawback.

Oxygen Production

Phytoplankton absorb energy from the Sun and nutrients from the water to produce their own nourishment or energy. In the process of photosynthesis, phytoplankton release molecular oxygen (O_2) into the water as a waste biproduct. It is estimated that about 50% of the world's oxygen is produced via phytoplankton photosynthesis. The rest is produced via photosynthesis on land by plants. Furthermore, phytoplankton photosynthesis has controlled the atmospheric CO_2/O_2 balance since the early Precambrian Eon.

Biomass Variability

Amphipod with curved exoskeleton and two long and two short antennae.

The growth of phytoplankton populations is dependent on light levels and nutrient availability. The chief factor limiting growth varies from region to region in the world's oceans. On a broad scale, growth of phytoplankton in the oligotrophic tropical and subtropical gyres is generally limited by nutrient supply, while light often limits phytoplankton growth in subarctic gyres. Environmental variability at multiple scales influences the nutrient and light available for phytoplankton, and as these organisms form the base of the marine food web, this variability in phytoplankton growth influences higher trophic levels. For example, at interannual scales phytoplankton levels temporarily plummet during El Niño periods, influencing populations of zooplankton, fishes, sea birds, and marine mammals.

The effects of anthropogenic warming on the global population of phytoplankton is an area of active research. Changes in the vertical stratification of the water column, the rate of temperature-dependent biological reactions, and the atmospheric supply of nutrients are expected to have important impacts on future phytoplankton productivity. Additionally, changes in the mortality of phytoplankton due to rates of zooplankton grazing may be significant.

Freshly hatched fish larvae are also plankton for a few days, as long as it takes before they can swim against currents.

Importance to Fish

Zooplankton are the initial prey item for almost all fish larvae as they switch from their yolk sacs to external feeding. Fish rely on the density and distribution of zooplankton to match that of new larvae, which can otherwise starve. Natural factors (e.g., current variations) and man-made factors (e.g. river dams) can strongly affect zooplankton, which can in turn strongly affect larval survival, and therefore breeding success.

The importance of both phytoplankton and zooplankton is also well-recognized in extensive and semi-intensive pond fish farming. Plankton population based pond management strategies for fish rearing have been practised by traditional fish farmers for decades, illustrating the importance of plankton even in man-made environments.

Phytoplankton

Phytoplankton are the autotrophic (self-feeding) components of the plankton community and a key part of oceans, seas and freshwater basin ecosystems. Most phytoplankton are too small to be individually seen with the unaided eye. However, when present in high enough numbers, some varieties may be noticeable as colored patches on the water surface due to the presence of chlorophyll within their cells and accessory pigments (such as phycobiliproteins or xanthophylls) in some species. About 1% of the global biomass is due to phytoplankton,

Types

Phytoplankton are extremely diverse, varying from photosynthesising bacteria (cyanobacteria), to plant-like diatoms, to armour-plated coccolithophores.

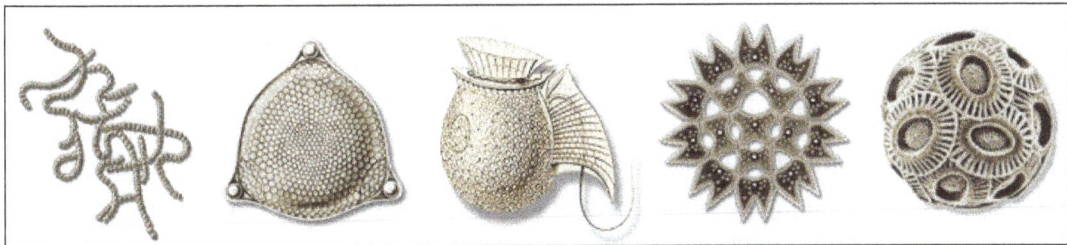

Cyanobacteria, diatom, dinoflagellate, greenalgae, coccolithophore, (drawings not to scale).

Ecology

Phytoplankton come in many shapes and sizes.

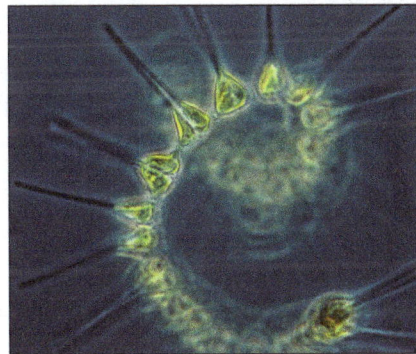

Phytoplankton are the foundation of the oceanic food chain.

Carbon

Phytoplankton are photosynthesizing microscopic biotic organisms that inhabit the upper sunlit layer of almost all oceans and bodies of fresh water on Earth. They are agents for "primary production", the creation of organic compounds from carbon dioxide dissolved in the water, a process that sustains the aquatic food web.

Phytoplankton obtain energy through the process of photosynthesis and must therefore live in the well-lit surface layer (termed the euphotic zone) of an ocean, sea, lake, or other body of water. Phytoplankton account for about half of all photosynthetic activity on Earth. Their cumulative energy fixation in carbon compounds (primary production) is the basis for the vast majority of oceanic and also many freshwater food webs (chemosynthesis is a notable exception).

While almost all phytoplankton species are obligate photoautotrophs, there are some that are mixotrophic and other, non-pigmented species that are actually heterotrophic (the latter are often viewed as zooplankton). Of these, the best known are dinoflagellate genera such as *Noctiluca* and *Dinophysis*, that obtain organic carbon by ingesting other organisms or detrital material.

Oxygen Production

Phytoplankton absorb energy from the Sun and nutrients from the water to produce their own food. In the process of photosynthesis, phytoplankton release molecular oxygen (O_2) into the water. It is estimated that between 50% and 85% of the world's oxygen is produced via phytoplankton photosynthesis. The rest is produced via photosynthesis on land by plants. Furthermore, phytoplankton photosynthesis has controlled the atmospheric CO_2/O_2 balance since the early Precambrian Eon.

Minerals

Phytoplankton are crucially dependent on minerals. These are primarily macronutrients such as nitrate, phosphate or silicic acid, whose availability is governed by the balance between the so-called biological pump and upwelling of deep, nutrient-rich waters. Phytoplankton nutrient composition drives and is driven by the Redfield ratio of macronutrients generally available throughout the surface oceans. However, across large regions of the World Ocean such as the Southern Ocean, phytoplankton are also limited by the lack of the micronutrient iron. This has led to some scientists advocating iron fertilization as a means to counteract the accumulation of human-produced carbon dioxide (CO_2) in the atmosphere. Large-scale experiments have added iron (usually as salts such as iron sulphate) to the oceans to promote phytoplankton growth and draw atmospheric CO_2 into the ocean. However, controversy about manipulating the ecosystem and the efficiency of iron fertilization has slowed such experiments.

B Vitamins

Phytoplankton depend on B Vitamins for survival. Areas in the ocean have been identified as having a major lack of some B Vitamins, and correspondingly, phytoplankton.

Temperature

The effects of anthropogenic warming on the global population of phytoplankton is an area of active research. Changes in the vertical stratification of the water column, the rate of temperature-dependent biological reactions, and the atmospheric supply of nutrients are expected to have important effects on future phytoplankton productivity.

PH

The effects of anthropogenic ocean acidification on phytoplankton growth and community structure has also received considerable attention. Phytoplankton such as coccolithophores contain calcium carbonate cell walls that are sensitive to ocean acidification. Because of their short generation times, evidence suggests some phytoplankton can adapt to changes in pH induced by increased carbon dioxide on rapid time-scales (months to years).

Food Web

Phytoplankton serve as the base of the aquatic food web, providing an essential ecological function for all aquatic life. Under future conditions of anthropogenic warming and ocean acidification, changes in phytoplankton mortality may be significant. One of the many food chains in the ocean – remarkable due to the small number of links – is that of phytoplankton sustaining krill (a crustacean similar to a tiny shrimp), which in turn sustain baleen whales.

Structural and Functional Diversity

The term phytoplankton encompasses all photoautotrophic microorganisms in aquatic food webs. However, unlike terrestrial communities, where most autotrophs are plants, phytoplankton are a diverse group, incorporating protistan eukaryotes and both eubacterial and archaebacterial prokaryotes. There are about 5,000 known species of marine phytoplankton. How such diversity evolved despite scarce resources (restricting niche differentiation) is unclear.

In terms of numbers, the most important groups of phytoplankton include the diatoms, cyanobacteria and dinoflagellates, although many other groups of algae are represented. One group, the coccolithophorids, is responsible (in part) for the release of significant amounts of dimethyl sulfide (DMS) into the atmosphere. DMS is oxidized to form sulfate which, in areas where ambient aerosol particle concentrations are low, can contribute to the population of cloud condensation nuclei, mostly leading to increased cloud cover and cloud albedo according to the so-called CLAW Hypothesis. Different types of phytoplankton support different trophic levels within varying ecosystems. In oligotrophic oceanic regions such as the Sargasso Sea or the South Pacific Gyre, phytoplankton is dominated by the small sized cells, called picoplankton and nanoplankton (also referred to as picoflagellates and nanoflagellates), mostly composed of cyanobacteria (*Prochlorococcus*, *Synechococcus*) and picoeucaryotes such as *Micromonas*. Within more productive ecosystems, dominated by upwelling or high terrestrial inputs, larger dinoflagellates are the more dominant phytoplankton and reflect a larger portion of the biomass.

Growth Strategy

In the early twentieth century, Alfred C. Redfield found the similarity of the phytoplankton's elemental composition to the major dissolved nutrients in the deep ocean. Redfield proposed that the ratio of carbon to nitrogen to phosphorus (106:16:1) in the ocean was controlled by the phytoplankton's requirements, as phytoplankton subsequently release nitrogen and phosphorus as they are remineralized. This so-called "Redfield ratio" in describing stoichiometry of phytoplankton and seawater has become a fundamental principle to understand marine ecology, biogeochemistry and phytoplankton evolution. However, the Redfield ratio is not a universal value and it may diverge due to the changes in exogenous nutrient delivery and microbial metabolisms in the ocean, such as nitrogen fixation, denitrification and anammox.

The dynamic stoichiometry shown in unicellular algae reflects their capability to store nutrients in an internal pool, shift between enzymes with various nutrient requirements and alter osmolyte composition. Different cellular components have their own unique stoichiometry characteristics, for instance, resource (light or nutrients) acquisition machinery such as proteins and chlorophyll contain a high concentration of nitrogen but low in phosphorus. Meanwhile, growth machinery such as ribosomal RNA contains high nitrogen and phosphorus concentrations.

Based on allocation of resources, phytoplankton is classified into three different growth strategies, namely survivalist, bloomer and generalist. Survivalist phytoplankton has a high ratio of N:P (>30) and contains an abundance of resource-acquisition machinery to sustain growth under scarce resources. Bloomer phytoplankton has a low N:P ratio (<10), contains a high proportion of growth machinery, and is adapted to exponential growth. Generalist phytoplankton has similar N:P to the Redfield ratio and contain relatively equal resource-acquisition and growth machinery.

Zooplankton

Zooplankton are heterotrophic (sometimes detritivorous) plankton (cf. phytoplankton). Plankton are organisms drifting in oceans, seas, and bodies of fresh water. Individual zooplankton are usually microscopic, but some (such as jellyfish) are larger and visible to the naked eye.

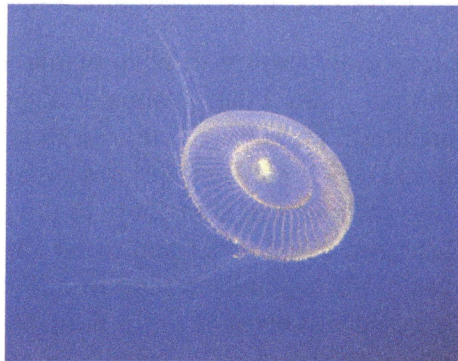

A jellyfish (Aequorea victoria).

Zooplankton is a categorization spanning a range of organism sizes including small protozoans and large metazoans. It includes holoplanktonic organisms whose complete life cycle lies within the plankton, as well as meroplanktonic organisms that spend part of their lives in the plankton before graduating to either the nekton or a sessile, benthic existence. Although zooplankton are primarily transported by ambient water currents, many have locomotion, used to avoid predators (as in diel vertical migration) or to increase prey encounter rate.

Ecologically important protozoan zooplankton groups include the foraminiferans, radiolarians and dinoflagellates (the last of these are often mixotrophic). Important metazoan zooplankton include cnidarians such as jellyfish and the Portuguese Man o' War; crustaceans such as copepods, ostracods, isopods, amphipods, mysids and krill; chaetognaths (arrow worms); molluscs such as pteropods; and chordates such as salps and juvenile fish. This wide phylogenetic range includes a similarly wide range in feeding behavior: filter feeding, predation and symbiosis with autotrophic phytoplankton as seen in corals. Zooplankton feed on bacterioplankton, phytoplankton, other zooplankton (sometimes cannibalistically), detritus (or marine snow) and even nektonic organisms. As a result, zooplankton are primarily found in surface waters where food resources (phytoplankton or other zooplankton) are abundant.

Just as any species can be limited within a geographical region, so are zooplankton. However, species of zooplankton are not dispersed uniformly or randomly within a region of the ocean. As with phytoplankton, 'patches' of zooplankton species exist throughout the ocean. Though few physical barriers exist above the mesopelagic, specific species of zooplankton are strictly restricted by salinity and temperature gradients; while other species can withstand wide temperature and salinity gradients. Zooplankton patchiness can also be influenced by biological factors, as well as other physical factors. Biological factors include breeding, predation, concentration of phytoplankton, and vertical migration. The physical factor that influences zooplankton distribution the most is mixing of the water column (upwelling and downwelling along the coast and in the open ocean) that affects nutrient availability and, in turn, phytoplankton production.

Through their consumption and processing of phytoplankton and other food sources, zooplankton play a role in aquatic food webs, as a resource for consumers on higher trophic levels (including fish), and as a conduit for packaging the organic material in the biological pump. Since they are typically small, zooplankton can respond rapidly to increases in phytoplankton abundance, for instance, during the spring bloom.

Zooplankton can also act as a disease reservoir. Crustacean zooplankton have been found to house the bacterium *Vibrio cholerae*, which causes cholera, by allowing the cholera vibrios to attach to their chitinous exoskeletons. This symbiotic relationship enhances the bacterium's ability to survive in an aquatic environment, as the exoskeleton provides the bacterium with carbon and nitrogen.

References

- Types-of-Marine-Organisms-HS-Es, lesson, types-of-marine-organisms, earth-science; ck12.org, Retrieved 18 January, 2019

- Benthos, science: britannica.com, Retrieved 19 February, 2019

- Walag, Angelo Mark; Mae Oljae P. Canencia (2016). "Physico-chemical parameters and macrobenthic invertebrates of the intertidal zone of Gusa, Cagayan de Oro City, Philippines" (PDF). Advances in Environmental Sciences. 8 (1): 71–82. Retrieved October 27, 2015

- Giere, Olav (2009). Meiobenthology. The microscopic motile fauna of aquatic sediments, 2nd edition, Springer. ISBN 978-3-540-68657-6

- Nekton, science: britannica.com, Retrieved 20 March, 2019

- Docker, Margaret F. (2006). "Bill Beamish's Contributions to Lamprey Research and Recent Advances in the Field". Guelph Ichthyology Reviews. 7. Doi:10.1111/j.1095-8649.2006.00968.x

- Rosenberg, Gary (2014). "A New Critical Estimate of Named Species Level Diversity of the Recent Mollusca". American Malacological Bulletin. 32 (2): 308–322. Doi:10.4003/006.032.0204

- "Cephalothorax". Crustacean Glossary. Natural History Museum of Los Angeles County. Retrieved 2016-09-10

- Chapman, A.D. (2009). Numbers of Living Species in Australia and the World, 2nd edition. Australian Biological Resources Study, Canberra. Retrieved 2010-01-12. ISBN 978-0-642-56860-1 (printed); ISBN 978-0-642-56861-8 (online)

- Bidle KD, Falkowski PG (August 2004). "Cell death in planktonic, photosynthetic microorganisms". Nature Reviews. Microbiology. 2 (8): 643–655. Doi:10.1038/nrmicro956. PMID 15263899

Marine Ecosystem

3

- **Marine Habitats**
- **Coral Reef**
- **Pelagic Zone**
- **Intertidal Zone**

Marine ecosystem is the Earth's largest aquatic ecosystem and is characterized by a high salt content. Some of the areas of study closely associated with marine ecosystems are marine habitats, coral reefs and intertidal zones. This chapter has been carefully written to provide an easy understanding of these facets of marine ecosystem.

Marine ecosystem is complex of living organisms in the ocean environment.

Marine waters cover two-thirds of the surface of the Earth. In some places the ocean is deeper than Mount Everest is high; for example, the Mariana Trench and the Tonga Trench in the western part of the Pacific Ocean reach depths in excess of 10,000 metres (32,800 feet). Within this ocean habitat live a wide variety of organisms that have evolved in response to various features of their environs.

Origins of Marine Life

The Earth formed approximately 4.5 billion years ago. As it cooled, water in the atmosphere condensed and the Earth was pummeled with torrential rains, which filled its great basins, forming seas. The primeval atmosphere and waters harboured the inorganic components hydrogen, methane, ammonia, and water. These substances are thought to have combined to form the first organic compounds when sparked by electrical discharges of lightning. Some of the earliest known organisms are cyanobacteria (formerly referred to as blue-green algae). Evidence of these early photosynthetic prokaryotes has been found in Australia in Precambrian marine sediments called stromatolites that are approximately 3 billion years old. Although the diversity of life-forms observed in modern oceans did not appear until much later, during the Precambrian (about 4.6 billion to 542 million years ago) many kinds of bacteria, algae, protozoa, and primitive metazoa evolved to exploit the early marine habitats of the world. During the Cambrian Period (about 542 million to 488 million years ago) a major radiation of life occurred in the oceans. Fossils of familiar organisms such as cnidaria (e.g., jellyfish), echinoderms (e.g., feather stars), precursors of the fishes (e.g., the protochordate Pikaia from the Burgess Shale of Canada), and other vertebrates are found in marine sediments of this age. The first fossil fishes are found in sediments from the Ordovician Period (about 488 million to 444 million years ago). Changes in the physical conditions of the ocean that are thought to have occurred in the Precambrian—an increase in the concentration of oxygen in seawater and a buildup of the ozone layer that reduced dangerous ultraviolet radiation—may have facilitated the increase and dispersal of living things.

Marine Environment

Geography, Oceanography and Topography

The shape of the oceans and seas of the world has changed significantly throughout the past 600 million years. According to the theory of plate tectonics, the crust of the Earth is made up of many dynamic plates. There are two types of plates—oceanic and continental—which float on the surface of the Earth's mantle, diverging, converging, or sliding against one another. When two plates diverge, magma from the mantle wells up and cools, forming new crust; when convergence occurs, one plate descends—i.e., is subducted—below the other and crust is resorbed into the mantle. Examples of both processes are observed in the marine environment. Oceanic crust is created along oceanic ridges or rift areas, which are vast undersea mountain ranges such as the Mid-Atlantic Ridge. Excess crust is reabsorbed along subduction zones, which usually are marked by deep-sea trenches such as the Kuril Trench off the coast of Japan.

The shape of the ocean also is altered as sea levels change. During ice ages a higher proportion of the waters of the Earth is bound in the polar ice caps, resulting in a relatively low sea level. When the polar ice caps melt during interglacial periods, the sea level rises. These changes in sea level cause great changes in the distribution of marine environments such as coral reefs. For example, during the last Pleistocene Ice Age the Great Barrier Reef did not exist as it does today; the continental shelf on which the reef now is found was above the high-tide mark.

Marine organisms are not distributed evenly throughout the oceans. Variations in characteristics of the marine environment create different habitats and influence what types of organisms will inhabit them. The availability of light, water depth, proximity to land, and topographic complexity all affect marine habitats.

The availability of light affects which organisms can inhabit a certain area of a marine ecosystem. The greater the depth of the water, the less light can penetrate until below a certain depth there is no light whatsoever. This area of inky darkness, which occupies the great bulk of the ocean, is called the aphotic zone. The illuminated region above it is called the photic zone, within which are distinguished the euphotic and disphotic zones. The euphotic zone is the layer closer to the surface that receives enough light for photosynthesis to occur. Beneath lies the disphotic zone, which is illuminated but so poorly that rates of respiration exceed those of photosynthesis. The actual depth of these zones depends on local conditions of cloud cover, water turbidity, and ocean surface. In general, the euphotic zone can extend to depths of 80 to 100 metres and the disphotic zone to depths of 80 to 700 metres. Marine organisms are particularly abundant in the photic zone, especially the euphotic portion; however, many organisms inhabit the aphotic zone and migrate vertically to the photic zone every night. Other organisms, such as the tripod fish and some species of sea cucumbers and brittle stars, remain in darkness all their lives.

Marine environments can be characterized broadly as a water, or pelagic, environment and a bottom, or benthic, environment. Within the pelagic environment the waters are divided into the neritic province, which includes the water above the continental shelf, and the oceanic province, which includes all the open waters beyond the continental shelf. The high nutrient levels of the neritic province—resulting from dissolved materials in riverine runoff—distinguish this province from the oceanic. The upper portion of both the neritic and oceanic waters—the epipelagic zone—is where photosynthesis occurs; it is roughly equivalent to the photic zone. Below this zone lie the mesopelagic, ranging between 200 and 1,000 metres, the bathypelagic, from 1,000 to 4,000 metres, and the abyssalpelagic, which encompasses the deepest parts of the oceans from 4,000 metres to the recesses of the deep-sea trenches.

The benthic environment also is divided into different zones. The supralittoral is above the high-tide mark and is usually not under water. The intertidal, or littoral, zone ranges from the high-tide mark (the maximum elevation of the tide) to the shallow, offshore waters. The sublittoral is the environment beyond the low-tide mark and is often used to refer to substrata of the continental shelf, which reaches depths of between 150 and 300 metres. Sediments of the continental shelf that influence marine organisms generally originate from the land, particularly in the form of riverine runoff, and include clay, silt, and sand. Beyond the continental shelf is the bathyal zone, which occurs at depths of 150 to 4,000 metres and includes the descending continental slope and rise. The abyssal zone (between 4,000 and 6,000 metres) represents a substantial portion of the oceans. The deepest region of the oceans (greater than 6,000 metres) is the hadal zone of the deep-sea trenches. Sediments of the deep sea primarily originate from a rain of dead marine organisms and their wastes.

Physical and Chemical Properties of Seawater

The physical and chemical properties of seawater vary according to latitude, depth, nearness to land, and input of fresh water. Approximately 3.5 percent of seawater is composed of dissolved compounds, while the other 96.5 percent is pure water. The chemical composition of seawater reflects such processes as erosion of rock and sediments, volcanic activity, gas exchange with the atmosphere, the metabolic and breakdown products of organisms, and rain. (For a list of the principal constituents of seawater, see seawater: Dissolved inorganic substances.) In addition to carbon, the nutrients essential for living organisms include nitrogen and phosphorus, which are minor constituents of seawater and thus are often limiting factors in organic cycles of the ocean. Concentrations of phosphorus and nitrogen are generally low in the photic zone because they are rapidly taken up by marine organisms. The highest concentrations of these nutrients generally are found below 500 metres, a result of the decay of organisms. Other important elements include silicon and calcium (essential in the skeletons of many organisms such as fish and corals).'

The chemical composition of the atmosphere also affects that of the ocean. For example, carbon dioxide is absorbed by the ocean and oxygen is released to the atmosphere through the activities of marine plants. The dumping of pollutants into the sea also can affect the chemical makeup of the ocean, contrary to earlier assumptions that, for example, toxins could be safely disposed of there.

The physical and chemical properties of seawater have a great effect on organisms, varying especially with the size of the creature. As an example, seawater is viscous to very small animals (less than 1 millimetre [0.039 inch] long) such as ciliates but not to large marine creatures such as tuna.

Marine organisms have evolved a wide variety of unique physiological and morphological features that allow them to live in the sea. Notothenid fishes in Antarctica are able to inhabit waters as cold as −2 °C (28 °F) because of proteins in their blood that act as antifreeze. Many organisms are able to achieve neutral buoyancy by secreting gas into internal chambers, as cephalopods do, or into swim bladders, as some fish do; other organisms use lipids, which are less dense than water, to achieve this effect. Some animals, especially those in the aphotic zone, generate light to attract prey. Animals in the disphotic zone such as hatchetfish produce light by means of organs called photophores to break up the silhouette of their bodies and avoid visual detection by predators. Many marine animals can detect vibrations or sound in the water over great distances by means of specialized organs. Certain fishes have lateral line systems, which they use to detect prey, and whales have a sound-producing organ called a melon with which they communicate. Tolerance to differences in salinity varies greatly: stenohaline organisms have a low tolerance to salinity changes, whereas euryhaline organisms, which are found in areas where river and sea meet (estuaries), are very tolerant of large changes in salinity. Euryhaline organisms are also very tolerant of changes in temperature. Animals that migrate between fresh water and salt water, such as salmon or eels, are capable of controlling their osmotic environment by active pumping or the retention of salts. Body architecture varies greatly in marine waters. The body shape of the cnidarian by-the-wind-sailor (Velella velella)—an animal that lives on the surface of the water (pleuston) and sails with the assistance of a modified flotation chamber—contrasts sharply with the sleek, elongated shape of the barracuda.

Ocean Currents

The movements of ocean waters are influenced by numerous factors, including the rotation of the Earth (which is responsible for the Coriolis effect), atmospheric circulation patterns that influence surface waters, and temperature and salinity gradients between the tropics and the polar regions (thermohaline circulation). The resultant patterns of circulation range from those that cover great areas, such as the North Subtropical Gyre, which follows a path thousands of kilometres long, to small-scale turbulences of less than one metre.

Marine organisms of all sizes are influenced by these patterns, which can determine the range of a species. For example, krill (Euphausia superba) are restricted to the Antarctic Circumpolar Current. Distribution patterns of both large and small pelagic organisms are affected as well. Mainstream currents such as the Gulf Stream and East Australian Current transport larvae great distances. As a result cold temperate coral reefs receive a tropical infusion when fish and invertebrate larvae from the tropics are relocated to high latitudes by these currents. The successful recruitment of eels to Europe depends on the strength of the Gulf Stream to transport them from spawning sites in the Caribbean. Areas where the ocean is affected by nearshore features, such as estuaries, or areas in which there is a vertical salinity gradient (halocline) often exhibit intense biological activity. In these environments, small organisms can become concentrated, providing a rich supply of food for other animals.

Marine Habitats

Although it may seem like the ocean is all the same, there are many different habitats based on temperature, salinity, pressure, light, currents, and other factors. Organisms have adapted to these conditions in many interesting and effective ways. Covering 70% of Earth's surface, the oceans are home to a large portion of all life on Earth.

Marine habitats are habitats that support marine life. Marine life depends in some way on the salt-water that is in the sea. A habitat is an ecological or environmental area inhabited by one or more living species. The marine environment supports many kinds of these habitats.

Marine habitats can be divided into coastal and open ocean habitats. Coastal habitats are found in the area that extends from as far as the tide comes in on the shoreline out to the edge of the continental shelf. Most marine life is found in coastal habitats, even though the shelf area occupies only seven percent of the total ocean area. Open ocean habitats are found in the deep ocean beyond the edge of the continental shelf.

Alternatively, marine habitats can be divided into pelagic and demersal zones. Pelagic habitats are found near the surface or in the open water column, away from the bottom of the ocean. Demersal habitats are near or on the bottom of the ocean. An organism living in a pelagic habitat is said to be a pelagic organism, as in pelagic fish. Similarly, an organism living in a demersal habitat is said to be a demersal organism, as in demersal fish. Pelagic habitats are intrinsically shifting and ephemeral, depending on what ocean currents are doing.

Marine habitats can be modified by their inhabitants. Some marine organisms, like corals, kelp, mangroves and seagrasses, are ecosystem engineers which reshape the marine environment to the point where they create further habitat for other organisms. By volume, oceans provide about 99 percent of the living space on the planet.

Only 29 percent of the world surface is land. The rest is ocean, home to the marine habitats. The oceans are nearly four kilometres deep on average and are fringed with coastlines that run for nearly 380,000 kilometres.

In contrast to terrestrial habitats, marine habitats are shifting and ephemeral. Swimming organisms find areas by the edge of a continental shelf a good habitat, but only while upwellings bring nutrient rich water to the surface. Shellfish find habitat on sandy beaches, but storms, tides and currents mean their habitat continually reinvents itself.

The presence of seawater is common to all marine habitats. Beyond that many other things determine whether a marine area makes a good habitat and the type of habitat it makes. For example:

- Temperature: Is affected by geographical latitude, ocean currents, weather, the discharge of rivers, and by the presence of hydrothermal vents or cold seeps.

- Sunlight: Photosynthetic processes depend on how deep and turbid the water is.

- Nutrients: Are transported by ocean currents to different marine habitats from land runoff, or by upwellings from the deep sea, or they sink through the sea as marine snow.

- Salinity:Varies, particularly in estuaries or near river deltas, or by hydrothermal vents.

- Dissolved gases: Oxygen levels in particular, can be increased by wave actions and decreased during algal blooms.

- Acidity: This is partly to do with dissolved gases above, since the acidity of the ocean is largely controlled by how much carbon dioxide is in the water.

- Turbulence: Ocean waves, fast currents and the agitation of water affect the nature of habitats.

- Cover: The availability of cover such as the adjacency of the sea bottom, or the presence of floating objects.

- The occupying organisms themselves: Since organisms modify their habitats by the act of occupying them, and some, like corals, kelp, mangroves and seagrasses, create further habitats for other organisms.

There are five major oceans, of which the Pacific Ocean is nearly as large as the rest put together. Coastlines fringe the land for nearly 380,000 kilometres.

Ocean	Area million km²	%	Volume million cu km	%	Mean depth km	Max depth km	Coastline km	Ref
Pacific Ocean	155.6	46.4	679.6	49.6	4.37	10.924	135,663	
Atlantic Ocean	76.8	22.9	313.4	22.5	4.08	8.605	111,866	
Indian Ocean	68.6	20.4	269.3	19.6	3.93	7.258	66,526	
Southern Ocean	20.3	6.1	91.5	6.7	4.51	7.235	17,968	
Arctic Ocean	14.1	4.2	17.0	1.2	1.21	4.665	45,389	
Overall	335.3		1370.8		4.09	10.924	377,412	

Land runoff, pouring into the sea, can contain nutrients.

Altogether, the ocean occupies 71 percent of the world surface, averaging nearly four kilometres in depth. By volume, the ocean provides about 99 percent of the living space on the planet. The science fiction writer Arthur C. Clarke has pointed out it would be more appropriate to refer to the planet Earth as the planet Sea or the planet Ocean.

Marine habitats can be broadly divided into pelagic and demersal habitats. Pelagic habitats are the habitats of the open water column, away from the bottom of the ocean. Demersal habitats are the habitats that are near or on the bottom of the ocean. An organism living in a pelagic habitat is said to be a pelagic organism, as in pelagic fish. Similarly, an organism living in a demersal habitat is said to be a demersal organism, as in demersal fish. Pelagic habitats are intrinsically ephemeral, depending on what ocean currents are doing.

The land-based ecosystem depends on topsoil and fresh water, while the marine ecosystem depends on dissolved nutrients washed down from the land. Ocean deoxygenation poses a threat to marine habitats, due to the growth of low oxygen zones.

Ocean Currents

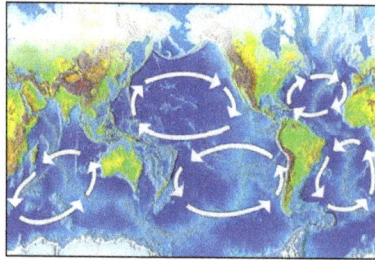

Ocean gyres rotate clockwise in the north and counterclockwise in the south.

In marine systems, ocean currents have a key role determining which areas are effective as habitats, since ocean currents transport the basic nutrients needed to support marine life. Plankton are the life forms that inhabit the ocean that are so small (less than 2 mm) that they cannot effectively propel themselves through the water, but must drift instead with the currents. If the current carries the right nutrients, and if it also flows at a suitably shallow depth where there is plenty of sunlight, then such a current itself can become a suitable habitat for photosynthesizing tiny algae called phytoplankton. These tiny plants are the primary producers in the ocean, at the start of the food chain. In turn, as the population of drifting phytoplankton grows, the water becomes a suitable habitat for zooplankton, which feed on the phytoplankton. While phytoplankton are tiny drifting plants, zooplankton are tiny drifting animals, such as the larvae of fish and marine invertebrates. If sufficient zooplankton establish themselves, the current becomes a candidate habitat for the forage fish that feed on them. And then if sufficient forage fish move to the area, it becomes a candidate habitat for larger predatory fish and other marine animals that feed on the forage fish. In this dynamic way, the current itself can, over time, become a moving habitat for multiple types of marine life.

This algae bloom occupies sunlit epipelagic waters off the southern coast of England. The algae are maybe feeding on nutrients from land runoff or upwellings at the edge of the continental shelf.

Ocean currents can be generated by differences in the density of the water. How dense water is depends on how saline or warm it is. If water contains differences in salt content or temperature, then the different densities will initiate a current. Water that is saltier or cooler will be denser, and will sink in relation to the surrounding water. Conversely, warmer and less salty water will float to the surface. Atmospheric winds and pressure differences also produces surface currents, waves and seiches. Ocean currents are also generated by the gravitational pull of the sun and moon (tides), and seismic activity (tsunami).

The rotation of the Earth affects the direction ocean currents take, and explains which way the large circular ocean gyres rotate in the image above left. Suppose a current at the equator is heading north. The Earth rotates eastward, so the water possesses that rotational momentum. But the further the water moves north, the slower the earth moves eastward. If the current could get to the North Pole, the earth wouldn't be moving eastward at all. To conserve its rotational momentum, the further the current travels north the faster it must move eastward. So the effect is that the current curves to the right. This is the Coriolis effect. It is weakest at the equator and strongest at the poles. The effect is opposite south of the equator, where currents curve left.

Marine Topography

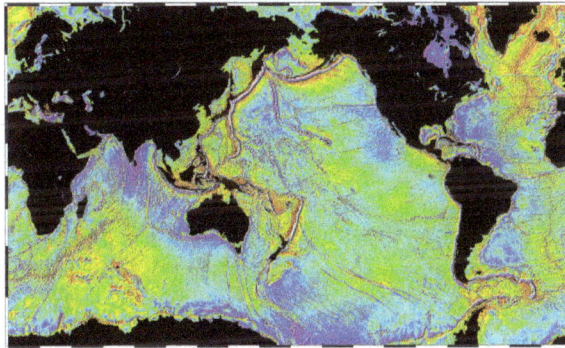

Map of underwater topography.

Marine (or seabed or ocean) topography refers to the shape the land has when it interfaces with the ocean. These shapes are obvious along coastlines, but they occur also in significant ways underwater. The effectiveness of marine habitats is partially defined by these shapes, including the way they interact with and shape ocean currents, and the way sunlight diminishes when these landforms occupy increasing depths. Tidal networks depend on the balance between sedimentary processes and hydrodynamics however, anthropogenic influences can impact the natural system more than any physical driver.

Marine topographies include coastal and oceanic landforms ranging from coastal estuaries and shorelines to continental shelves and coral reefs. Further out in the open ocean, they include underwater and deep sea features such as ocean rises and seamounts. The submerged surface has mountainous features, including a globe-spanning mid-ocean ridge system, as well as undersea volcanoes, oceanic trenches, submarine canyons, oceanic plateaus and abyssal plains.

The mass of the oceans is approximately 1.35×10^{18} metric tons, or about 1/4400 of the total mass of the Earth. The oceans cover an area of 3.618×10^8 km^2 with a mean depth of 3,682 m, resulting in an estimated volume of 1.332×10^9 km^3.

Biomass

One measure of the relative importance of different marine habitats is the rate at which they produce biomass.

Producer	Biomass productivity (gC/m²/yr)	Total area (million km²)	Total production (billion tonnes C/yr)	Comment
Swamps and marshes	2,500			Includes freshwater
Coral reefs	2,000	0.28	0.56	
Algal beds	2,000			
River estuaries	1,800			
Open ocean	125	311	39	

Coastal

Coastlines can be volatile habitats.

Marine coasts are dynamic environments which constantly change, like the ocean which partially shape them. The Earth's natural processes, including weather and sea level change, result in the erosion, accretion and resculpturing of coasts as well as the flooding and creation of continental shelves and drowned river valleys.

The main agents responsible for deposition and erosion along coastlines are waves, tides and currents. The formation of coasts also depends on the nature of the rocks they are made of: the harder the rocks the less likely they are to erode, so variations in rock hardness result in coastlines with different shapes.

Tides often determine the range over which sediment is deposited or eroded. Areas with high tidal ranges allow waves to reach farther up the shore, and areas with lower tidal ranges produce deposition at a smaller elevation interval. The tidal range is influenced by the size and shape of the coastline. Tides do not typically cause erosion by themselves; however, tidal bores can erode as the waves surge up river estuaries from the ocean.

Waves erode coastline as they break on shore releasing their energy; the larger the wave the more energy it releases and the more sediment it moves. Sediment deposited by waves comes from eroded cliff faces and is moved along the coastline by the waves. Sediment deposited by rivers is the dominant influence on the amount of sediment located on a coastline.

Shores that look permanent through the short perceptive of a human lifetime are in fact among the most temporary of all marine structures.

The sedimentologist Francis Shepard classified coasts as primary or secondary:

- Primary coasts are shaped by non-marine processes, by changes in the land form. If a coast is in much the same condition as it was when sea level was stabilised after the last ice age, it is called a primary coast. "Primary coasts are created by erosion (the wearing away of soil or rock), deposition (the buildup of sediment or sand) or tectonic activity (changes in the structure of the rock and soil because of earthquakes). Many of these coastlines were formed as the sea level rose during the last 18,000 years, submerging river and glacial valleys to form bays and fjords." An example of a primary coast is a river delta, which forms when a river deposits soil and other material as it enters the sea.

- Secondary coasts are produced by marine processes, such as the action of the sea or by creatures that live in it. Secondary coastlines include sea cliffs, barrier islands, mud flats, coral reefs, mangrove swamps and salt marshes.

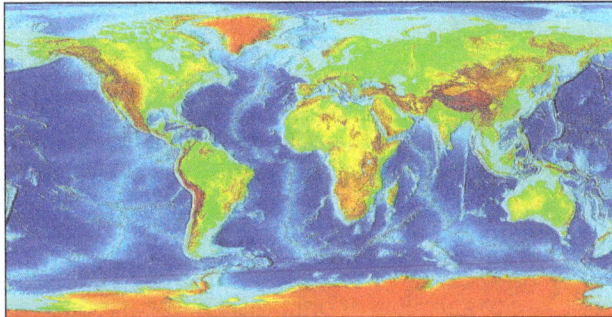

The global continental shelf, highlighted in cyan, defines the extent of coastal habitats, and occupies 5% of the total world area.

Continental coastlines usually have a continental shelf, a shelf of relatively shallow water, less than 200 metres deep, which extends 68 km on average beyond the coast. Worldwide, continental shelves occupy a total area of about 24 million km² (9 million sq mi), 8% of the ocean's total area and nearly 5% of the world's total area. Since the continental shelf is usually less than 200 metres deep, it follows that coastal habitats are generally photic, situated in the sunlit epipelagic zone. This means the conditions for photosynthetic processes so important for primary production, are available to coastal marine habitats. Because land is nearby, there are large discharges of nutrient rich land runoff into coastal waters. Further, periodic upwellings from the deep ocean can provide cool and nutrient rich currents along the edge of the continental shelf.

As a result, coastal marine life is the most abundant in the world. It is found in tidal pools, fjords and estuaries, near sandy shores and rocky coastlines, around coral reefs and on or above the continental shelf. Coastal fish include small forage fish as well as the larger predator fish that feed on them. Forage fish thrive in inshore waters where high productivity results from upwelling and shoreline run off of nutrients. Some are partial residents that spawn in streams, estuaries and bays, but most complete their life cycle in the zone. There can also be a mutualism between species that occupy adjacent marine habitats. For example, fringing reefs just below low tide level have a mutually beneficial relationship with mangrove forests at high tide level and sea grass meadows in between: the reefs protect the mangroves and seagrass from strong currents and waves that would damage them or erode the sediments in which they are rooted, while the mangroves and seagrass protect the coral from large influxes of silt, fresh water and pollutants. This additional level of variety in the environment is beneficial to many types of coral reef animals, which for example may feed in the sea grass and use the reefs for protection or breeding.

Coastal habitats are the most visible marine habitats, but they are not the only important marine habitats. Coastlines run for 380,000 kilometres, and the total volume of the ocean is 1,370 million cu km. This means that for each metre of coast, there is 3.6 cu km of ocean space available somewhere for marine habitats.

Waves and currents shape the intertidal shoreline, eroding the softer rocks and transporting and grading loose particles into shingles, sand or mud.

Intertidal

Intertidal zones, those areas close to shore, are constantly being exposed and covered by the ocean's tides. A huge array of life lives within this zone. Shore habitats range from the upper intertidal zones to the area where land vegetation takes prominence. It can be underwater anywhere from daily to very infrequently. Many species here are scavengers, living off of sea life that is washed up on the shore. Many land animals also make much use of the shore and intertidal habitats. A subgroup of organisms in this habitat bores and grinds exposed rock through the process of bio-erosion.

Sandy Shores

Sandy shores provide shifting homes to many species.

Sandy shores, also called beaches, are coastal shorelines where sand accumulates. Waves and currents shift the sand, continually building and eroding the shoreline. Longshore currents flow parallel to the beaches, making waves break obliquely on the sand. These currents transport large amounts of sand along coasts, forming spits, barrier islands and tombolos. Longshore currents also commonly create offshore bars, which give beaches some stability by reducing erosion.

Sandy shores are full of life, The grains of sand host diatoms, bacteria and other microscopic creatures. Some fish and turtles return to certain beaches and spawn eggs in the sand. Birds habitat beaches, like gulls, loons, sandpipers, terns and pelicans. Aquatic mammals, such sea lions, recuperate on them. Clams, periwinkles, crabs, shrimp, starfish and sea urchins are found on most beaches.

Sand is a sediment made from small grains or particles with diameters between about 60 μm and 2 mm. Mud is a sediment made from particles finer than sand. This small particle size means that mud particles tend to stick together, whereas sand particles do not. Mud is not easily shifted by waves and currents, and when it dries out, cakes into a solid. By contrast, sand is easily shifted by waves and currents, and when sand dries out it can be blown in the wind, accumulating into shifting sand dunes. Beyond the high tide mark, if the beach is low-lying, the wind can form rolling hills of sand dunes. Small dunes shift and reshape under the influence of the wind while larger dunes stabilise the sand with vegetation.

Ocean processes grade loose sediments to particle sizes other than sand, such as gravel or cobbles. Waves breaking on a beach can leave a berm, which is a raised ridge of coarser pebbles or sand, at the high tide mark. Shingle beaches are made of particles larger than sand, such as cobbles, or small stones. These beaches make poor habitats. Little life survives because the stones are churned and pounded together by waves and currents.

Rocky Shores

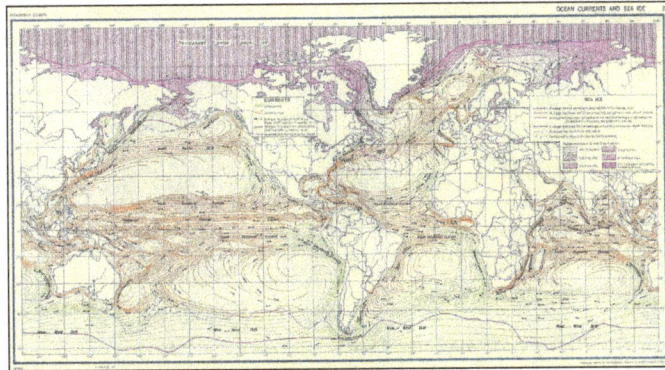

Tidepools on rocky shores make turbulent habitats for many forms of marine life.

The relative solidity of rocky shores seems to give them a permanence compared to the shifting nature of sandy shores. This apparent stability is not real over even quite short geological time scales, but it is real enough over the short life of an organism. In contrast to sandy shores, plants and animals can anchor themselves to the rocks.

Competition can develop for the rocky spaces. For example, barnacles can compete successfully on open intertidal rock faces to the point where the rock surface is covered with them. Barnacles resist desiccation and grip well to exposed rock faces. However, in the crevices of the same rocks, the inhabitants are different. Here mussels can be the successful species, secured to the rock with their byssal threads.

Rocky and sandy coasts are vulnerable because humans find them attractive and want to live near them. An increasing proportion of the humans live by the coast, putting pressure on coastal habitats.

Mudflats

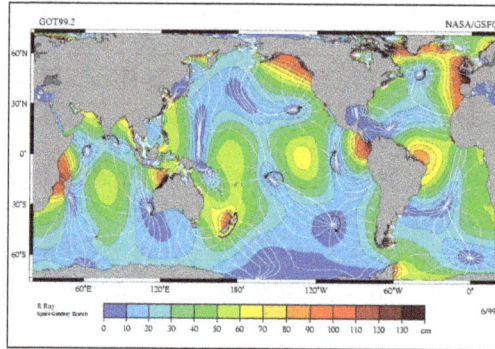

Mudflats become temporary habitats for migrating birds.

Mudflats are coastal wetlands that form when mud is deposited by tides or rivers. They are found in sheltered areas such as bays, bayous, lagoons, and estuaries. Mudflats may be viewed geologically as exposed layers of bay mud, resulting from deposition of estuarine silts, clays and marine animal detritus. Most of the sediment within a mudflat is within the intertidal zone, and thus the flat is submerged and exposed approximately twice daily.

Mudflats are typically important regions for wildlife, supporting a large population, although levels of biodiversity are not particularly high. They are of particular importance to migratory birds. In the United Kingdom mudflats have been classified as a Biodiversity Action Plan priority habitat.

Mangrove and Salt Marshes

Mangroves provide nurseries for fish.

Mangrove swamps and salt marshes form important coastal habitats in tropical and temperate areas respectively. Mangroves are species of shrubs and medium size trees that grow in saline coastal sediment habitats in the tropics and subtropics: mainly between latitudes 25° N and 25° S. The saline conditions tolerated by various species range from brackish water, through pure seawater (30 to 40 ppt), to water concentrated by evaporation to over twice the salinity of ocean seawater (up to 90 ppt). There are many mangrove species, not all closely related. The term "mangrove" is used generally to cover all of these species, and it can be used narrowly to cover just mangrove trees of the genus *Rhizophora*.

Mangroves form a distinct characteristic saline woodland or shrubland habitat, called a mangrove swamp or mangrove forest'. Mangrove swamps are found in depositional coastal environments, where fine sediments (often with high organic content) collect in areas protected from high-energy wave action. Mangroves dominate three quarters of tropical coastlines.

Estuaries

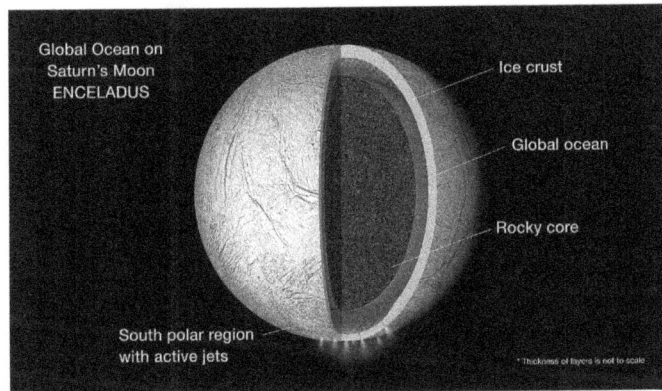

Estuaries occur when rivers flow into a coastal bay or inlet. They are nutrient rich and have a transition zone which moves from freshwater to saltwater.

An estuary is a partly enclosed coastal body of water with one or more rivers or streams flowing into it, and with a free connection to the open sea. Estuaries form a transition zone between river environments and ocean environments and are subject to both marine influences, such as tides, waves, and the influx of saline water; and riverine influences, such as flows of fresh water and sediment. The inflow of both seawater and freshwater provide high levels of nutrients in both the water column and sediment, making estuaries among the most productive natural habitats in the world.

Most estuaries were formed by the flooding of river-eroded or glacially scoured valleys when sea level began to rise about 10,000-12,000 years ago. They are amongst the most heavily populated areas throughout the world, with about 60% of the world's population living along estuaries and the coast. As a result, estuaries are suffering degradation by many factors, including sedimentation from soil erosion from deforestation; overgrazing and other poor farming practices; overfishing; drainage and filling of wetlands; eutrophication due to excessive nutrients from sewage and animal wastes; pollutants including heavy metals, PCBs, radionuclides and hydrocarbons from sewage inputs; and diking or damming for flood control or water diversion.

Estuaries provide habitats for a large number of organisms and support very high productivity. Estuaries provide habitats for salmon and sea trout nurseries, as well as migratory bird populations. Two of the main characteristics of estuarine life are the variability in salinity and sedimentation. Many species of fish and invertebrates have various methods to control or conform to the shifts in salt concentrations and are termed osmoconformers and osmoregulators. Many animals also burrow to avoid predation and to live in the more stable sedimental environment. However, large numbers of bacteria are found within the sediment which have a very high oxygen demand. This reduces the levels of oxygen within the sediment often resulting in partially anoxic conditions, which can be further exacerbated by limited water flux. Phytoplankton are key primary producers in estuaries. They move with the water bodies and can be flushed in and out with the tides. Their productivity is largely dependent on the turbidity of the water. The main phytoplankton present are diatoms and dinoflagellates which are abundant in the sediment.

Kelp Forests

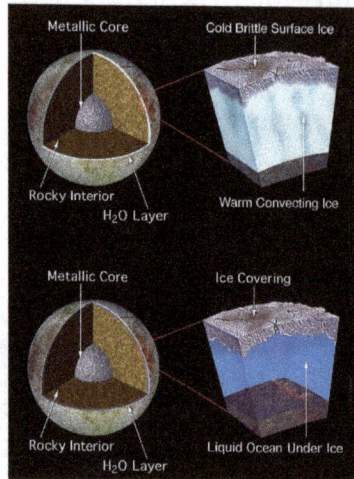

Kelp forests provide habitat for many marine organisms.

Kelp forests are underwater areas with a high density of kelp. They form some of the most productive and dynamic ecosystems on Earth. Smaller areas of anchored kelp are called *kelp beds*. Kelp forests occur worldwide throughout temperate and polar coastal oceans.

Kelp forests provide a unique three-dimensional habitat for marine organisms and are a source for understanding many ecological processes. Over the last century, they have been the focus of extensive research, particularly in trophic ecology, and continue to provoke important ideas that are relevant beyond this unique ecosystem. For example, kelp forests can influence coastal oceanographic patterns and provide many ecosystem services.

However, humans have contributed to kelp forest degradation. Of particular concern are the effects of overfishing nearshore ecosystems, which can release herbivores from their normal population regulation and result in the over-grazing of kelp and other algae. This can rapidly result in transitions to barren landscapes where relatively few species persist.

Frequently considered an ecosystem engineer, kelp provides a physical substrate and habitat for kelp forest communities. In algae (Kingdom: Protista), the body of an individual organism is known as a thallus rather than as a plant (Kingdom: Plantae). The morphological structure of a kelp thallus is defined by three basic structural units:

- The holdfast is a root-like mass that anchors the thallus to the sea floor, though unlike true roots it is not responsible for absorbing and delivering nutrients to the rest of the thallus;

- The stipe is analogous to a plant stalk, extending vertically from the holdfast and providing a support framework for other morphological features;

- The fronds are leaf- or blade-like attachments extending from the stipe, sometimes along its full length, and are the sites of nutrient uptake and photosynthetic activity.

In addition, many kelp species have pneumatocysts, or gas-filled bladders, usually located at the base of fronds near the stipe. These structures provide the necessary buoyancy for kelp to maintain an upright position in the water column.

The environmental factors necessary for kelp to survive include hard substrate (usually rock), high nutrients (e.g., nitrogen, phosphorus), and light (minimum annual irradiance dose > 50 E m^{-2}). Especially productive kelp forests tend to be associated with areas of significant oceanographic upwelling, a process that delivers cool nutrient-rich water from depth to the ocean's mixed surface layer. Water flow and turbulence facilitate nutrient assimilation across kelp fronds throughout the water column. Water clarity affects the depth to which sufficient light can be transmitted. In ideal conditions, giant kelp (*Macrocystis spp.*) can grow as much as 30-60 centimetres vertically per day. Some species such as *Nereocystis* are annual while others like *Eisenia* are perennial, living for more than 20 years. In perennial kelp forests, maximum growth rates occur during upwelling months (typically spring and summer) and die-backs correspond to reduced nutrient availability, shorter photoperiods and increased storm frequency.

Seagrass Meadows

White-spotted puffers like living in seagrass areas.

Seagrasses are flowering plants from one of four plant families which grow in marine environments. They are called *seagrasses* because the leaves are long and narrow and are very often green, and because the plants often grow in large meadows which look like grassland. Since seagrasses photosynthesize and are submerged, they must grow submerged in the photic zone, where there is enough sunlight. For this reason, most occur in shallow and sheltered coastal waters anchored in sand or mud bottoms.

Seagrasses form extensive beds or meadows, which can be either monospecific (made up of one species) or multispecific (where more than one species co-exist). Seagrass beds make highly diverse and productive ecosystems. They are home to phyla such as juvenile and adult fish, epiphytic and free-living macroalgae and microalgae, mollusks, bristle worms, and nematodes. Few species were originally considered to feed directly on seagrass leaves (partly because of their low nutritional content), but scientific reviews and improved working methods have shown that seagrass herbivory is a highly important link in the food chain, with hundreds of species feeding on seagrasses worldwide, including green turtles, dugongs, manatees, fish, geese, swans, sea urchins and crabs.

Seagrasses are ecosystem engineers in the sense that they partly create their own habitat. The leaves slow down water-currents increasing sedimentation, and the seagrass roots and rhizomes stabilize the seabed. Their importance to associated species is mainly due to provision of shelter (through their three-dimensional structure in the water column), and due to their extraordinarily high rate of primary production. As a result, seagrasses provide coastal zones with ecosystem services, such as fishing grounds, wave protection, oxygen production and protection against coastal erosion. Seagrass meadows account for 15% of the ocean's total carbon storage.

Coral Reefs

Reefs comprise some of the densest and most diverse habitats in the world. The best-known types of reefs are tropical coral reefs which exist in most tropical waters; however, reefs can also exist in cold water. Reefs are built up by corals and other calcium-depositing animals, usually on top of a rocky outcrop on the ocean floor. Reefs can also grow on other surfaces, which has made it possible to create artificial reefs. Coral reefs also support a huge community of life, including the corals themselves, their symbiotic zooxanthellae, tropical fish and many other organisms.

Much attention in marine biology is focused on coral reefs and the El Niño weather phenomenon. In 1998, coral reefs experienced the most severe mass bleaching events on record, when vast expanses of reefs across the world died because sea surface temperatures rose well above normal. Some reefs are recovering, but scientists say that between 50% and 70% of the world's coral reefs are now endangered and predict that global warming could exacerbate this trend.

Open Ocean

The open ocean is relatively unproductive because of a lack of nutrients, yet because it is so vast, it has more overall primary production than any other marine habitat. Only about 10 percent of marine species live in the open ocean. But among them are the largest and fastest of all marine animals, as well as the animals that dive the deepest and migrate the longest. In the depths lurk animal that, to our eyes, appear hugely alien.

Surface Waters

In the open ocean, sunlit surface epipelagic waters get enough light for photosynthesis,
but there are often not enough nutrients. As a result, large areas contain
little life apart from migrating animals.

The surface waters are sunlit. The waters down to about 200 metres are said to be in the epipelagic zone. Enough sunlight enters the epipelagic zone to allow photosynthesis by phytoplankton. The epipelagic zone is usually low in nutrients. This partially because the organic debris produced in the zone, such as excrement and dead animals, sink to the depths and are lost to the upper zone. Photosynthesis can happen only if both sunlight and nutrients are present.

In some places, like at the edge of continental shelves, nutrients can upwell from the ocean depth, or land runoff can be distributed by storms and ocean currents. In these areas, given that both sunlight and nutrients are now present, phytoplankton can rapidly establish itself, multiplying so fast that the water turns green from the chlorophyll, resulting in an algal bloom. These nutrient rich surface waters are among the most biologically productive in the world, supporting billions of tonnes of biomass.

"Phytoplankton are eaten by zooplankton - small animals which, like phytoplankton, drift in the ocean currents. The most abundant zooplankton species are copepods and krill: tiny crustaceans that are the most numerous animals on Earth. Other types of zooplankton include jelly fish and the larvae of fish, marine worms, starfish, and other marine organisms". In turn, the zooplankton are eaten by filter-feeding animals, including some seabirds, small forage fish like herrings and sardines, whale sharks, manta rays, and the largest animal in the world, the blue whale. Yet again, moving up the foodchain, the small forage fish are in turn eaten by larger predators, such as tuna, marlin, sharks, large squid, seabirds, dolphins, and toothed whales.

Deep Sea

The deep sea starts at the aphotic zone, the point where sunlight loses most of its energy in the water. Many life forms that live at these depths have the ability to create their own light a unique evolution known as bio-luminescence.

In the deep ocean, the waters extend far below the epipelagic zone, and support very different types of pelagic life forms adapted to living in these deeper zones.

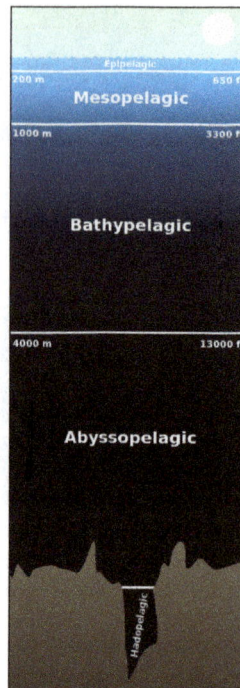
Scale diagram of the layers of the pelagic zone.

Much of the aphotic zone's energy is supplied by the open ocean in the form of detritus. In deep water, marine snow is a continuous shower of mostly organic detritus falling from the upper layers of the water column. Its origin lies in activities within the productive photic zone. Marine snow includes dead or dying plankton, protists (diatoms), fecal matter, sand, soot and other inorganic dust. The "snowflakes" grow over time and may reach several centimetres in diameter, travelling for weeks before reaching the ocean floor. However, most organic components of marine snow are consumed by microbes, zooplankton and other filter-feeding animals within the first 1,000 metres of their journey, that is, within the epipelagic zone. In this way marine snow may be considered the foundation of deep-sea mesopelagic and benthic ecosystems: As sunlight cannot reach them, deep-sea organisms rely heavily on marine snow as an energy source.

Some deep-sea pelagic groups, such as the lanternfish, ridgehead, marine hatchetfish, and lightfish families are sometimes termed *pseudoceanic* because, rather than having an even distribution in open water, they occur in significantly higher abundances around structural oases, notably seamounts and over continental slopes. The phenomenon is explained by the likewise abundance of prey species which are also attracted to the structures.

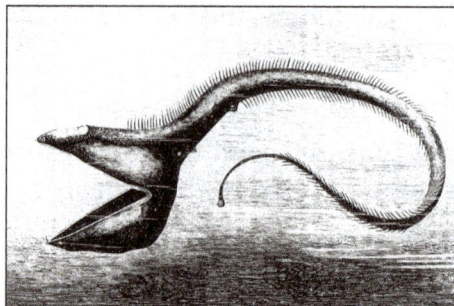
The umbrella mouth gulper eel can swallow a fish much larger than itself.

The fish in the different pelagic and deep water benthic zones are physically structured, and behave in ways, that differ markedly from each other. Groups of coexisting species within each zone all seem to operate in similar ways, such as the small mesopelagic vertically migrating plankton-feeders, the bathypelagic anglerfishes, and the deep water benthic rattails.

Ray finned species, with spiny fins, are rare among deep sea fishes, which suggests that deep sea fish are ancient and so well adapted to their environment that invasions by more modern fishes have been unsuccessful. The few ray fins that do exist are mainly in the Beryciformes and Lampriformes, which are also ancient forms. Most deep sea pelagic fishes belong to their own orders, suggesting a long evolution in deep sea environments. In contrast, deep water benthic species, are in orders that include many related shallow water fishes.

The umbrella mouth gulper is a deep sea eel with an enormous loosely hinged mouth. It can open its mouth wide enough to swallow a fish much larger than itself, and then expand its stomach to accommodate its catch.

Sea Floor

Vents and Seeps

Zooarium chimney provides a habitat for vent biota.

Hydrothermal vents along the mid-ocean ridge spreading centers act as oases, as do their opposites, cold seeps. Such places support unique marine biomes and many new marine microorganisms and other lifeforms have been discovered at these locations.

Trenches

The deepest recorded oceanic trenches measure to date is the Mariana Trench, near the Philippines, in the Pacific Ocean at 10,924 m (35,838 ft). At such depths, water pressure is extreme and there is no sunlight, but some life still exists. A white flatfish, a shrimp and a jellyfish were seen by the American crew of the bathyscaphe *Trieste* when it dove to the bottom in 1960.

Seamounts

Marine life also flourishes around seamounts that rise from the depths, where fish and other sea life congregate to spawn and feed.

Marine Life

Marine life, or sea life or ocean life, is the plants, animals and other organisms that live in the salt water of the sea or ocean, or the brackish water of coastal estuaries. At a fundamental level, marine life affects the nature of the planet. Marine organisms produce oxygen. Shorelines are in part shaped and protected by marine life, and some marine organisms even help create new land.

Most life forms evolved initially in marine habitats. By volume, oceans provide about 90 percent of the living space on the planet. The earliest vertebrates appeared in the form of fish, which live exclusively in water. Some of these evolved into amphibians which spend portions of their lives in water and portions on land. Other fish evolved into land mammals and subsequently returned to the ocean as seals, dolphins or whales. Plant forms such as kelp and algae grow in the water and are the basis for some underwater ecosystems. Plankton, and particularly phytoplankton, are key primary producers forming the general foundation of the ocean food chain.

Marine invertebrates exhibit a wide range of modifications to survive in poorly oxygenated waters, including breathing tubes as in mollusc siphons. Fish have gills instead of lungs, although some species of fish, such as the lungfish, have both. Marine mammals, such as dolphins, whales, otters, and seals need to surface periodically to breathe air.

A total of 230,000 documented marine species exist, including about 20,000 species of marine fish, with some two million marine species yet to be documented. Marine species range in size from the microscopic, including plankton and phytoplankton which can be as small as 0.02 micrometres, to huge cetaceans (whales, dolphins and porpoises), including the blue whale: the largest known animal reaching up to 33 metres (108 ft) in length. Marine microorganisms, including bacteria and viruses, constitute about 70% of the total marine biomass.

Water

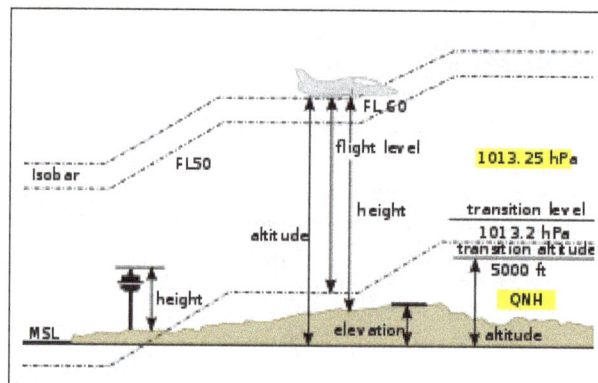

Elevation histogram showing the percentage of the Earth's surface above and below sea level.

There is no life without water. It has been described as the *universal solvent* for its ability to dissolve many substances, and as the *solvent of life*. Water is the only common substance to exist as a solid, liquid, and gas under conditions normal to life on Earth. The Nobel Prize winner Albert Szent-Györgyi referred to water as the *mater und matrix*: the mother and womb of life.

The abundance of surface water on Earth is a unique feature in the Solar System. Earth's hydrosphere consists chiefly of the oceans, but technically includes all water surfaces in the world, including inland seas, lakes, rivers, and underground waters down to a depth of 2,000 metres (6,600 ft) The deepest underwater location is Challenger Deep of the Mariana Trench in the Pacific Ocean, having a depth of 10,900 metres (6.8 mi).

The mass of the oceans is 1.35×10^{18} metric tons, or about 1/4400 of Earth's total mass. The oceans cover an area of 3.618×10^8 km^2 with a mean depth of 3682 m, resulting in an estimated volume of 1.332×10^9 km^3. If all of Earth's crustal surface was at the same elevation as a smooth sphere, the depth of the resulting world ocean would be about 2.7 kilometres (1.7 mi).

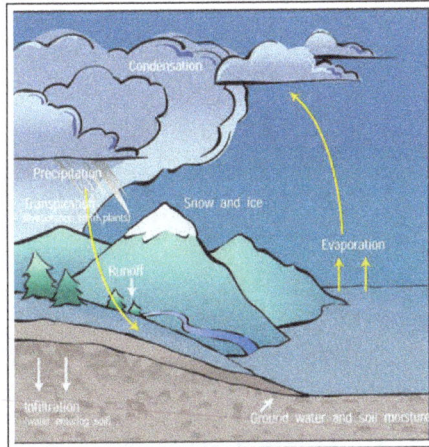

The Earth's water cycle.

About 97.5% of the water on Earth is saline; the remaining 2.5% is fresh water. Most fresh water: about 69%: is present as ice in ice caps and glaciers. The average salinity of Earth's oceans is about 35 grams (1.2 oz) of salt per kilogram of seawater (3.5% salt). Most of the salt in the ocean comes from the weathering and erosion of rocks on land. Some salts are released from volcanic activity or extracted from cool igneous rocks.

The oceans are also a reservoir of dissolved atmospheric gases, which are essential for the survival of many aquatic life forms. Sea water has an important influence on the world's climate, with the oceans acting as a large heat reservoir. Shifts in the oceanic temperature distribution can cause significant weather shifts, such as the El Niño-Southern Oscillation.

Jupiter's moon Europa may have an underground ocean which supports life.

Altogether the ocean occupies 71 percent of the world surface, averaging nearly 3.7 kilometres (2.3 mi) in depth. By volume, the ocean provides about 90 percent of the living space on the planet. The science fiction writer Arthur C. Clarke has pointed out it would be more appropriate to refer to planet Earth as planet Ocean.

However water is found elsewhere in the solar system. Europa, one of the moons orbiting Jupiter, is slightly smaller than the Earth's moon. There is a strong possibility a large saltwater ocean exists beneath its ice surface. It has been estimated the outer crust of solid ice is about 10–30 km (6–19 mi) thick and the liquid ocean underneath is about 100 km (60 mi) deep. This would make Europa's ocean over twice the volume of the Earth's ocean. There has been speculation Europa's ocean could support life, and could be capable of supporting multicellular microorganisms if hydrothermal vents are active on the ocean floor.

Microorganisms

Microbial Mats

Microbial mats are the earliest form of life on Earth for which there is good fossil evidence. The image shows a cyanobacterial-algal mat.

Stromatolites are formed from microbial mats as microbes slowly move upwards to avoid being smothered by sediment.

Microorganisms make up about 70% of the marine biomass. A microorganism, or microbe, is a microscopic organism too small to be recognised with the naked eye. It can be single-celled or multicellular. Microorganisms are diverse and include all bacteria and archaea, most protozoa such as algae, fungi and certain microscopic animals such as rotifers. Many macroscopic animals and plants have microscopic juvenile stages. Some microbiologists also classify viruses (and viroids) as microorganisms, but others consider these as nonliving.

Microorganisms are crucial to nutrient recycling in ecosystems as they act as decomposers. Some microorganisms are pathogenic, causing disease and even death in plants and animals. As inhabitants of the largest environment on Earth, microbial marine systems drive changes in every global system. Microbes are responsible for virtually all the photosynthesis that occurs in the ocean, as well as the cycling of carbon, nitrogen, phosphorus and other nutrients and trace elements.

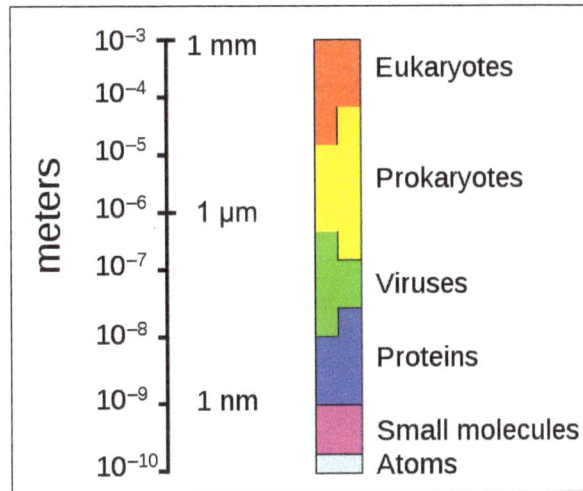

The range of sizes shown by prokaryotes (bacteria and archaea) and viruses relative to those of other organisms and biomolecules.

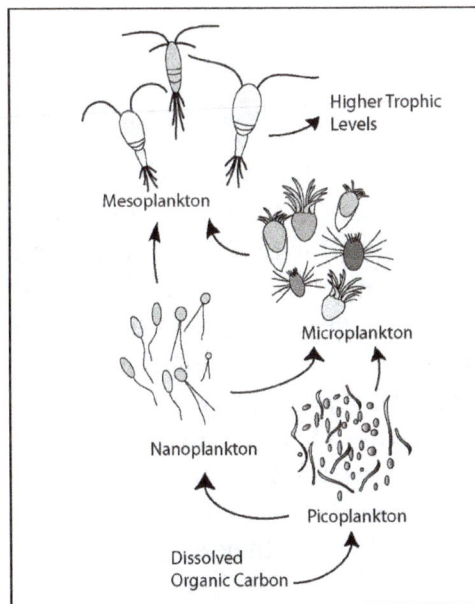

Marine microbial loop.

Microscopic life undersea is diverse and still poorly understood, such as for the role of viruses in marine ecosystems. Most marine viruses are bacteriophages, which are harmless to plants and animals, but are essential to the regulation of saltwater and freshwater ecosystems. They infect and destroy bacteria in aquatic microbial communities, and are the most important mechanism of recycling carbon in the marine environment. The organic molecules released from the dead bacterial cells stimulate fresh bacterial and algal growth. Viral activity may also contribute to the biological pump, the process whereby carbon is sequestered in the deep ocean.

A stream of airborne microorganisms circles the planet above weather systems but below commercial air lanes. Some peripatetic microorganisms are swept up from terrestrial dust storms, but most originate from marine microorganisms in sea spray. In 2018, scientists reported that hundreds of millions of viruses and tens of millions of bacteria are deposited daily on every square meter around the planet.

Sea spray containing marine microorganisms can be swept high into the atmosphere and may travel the globe before falling back to earth.

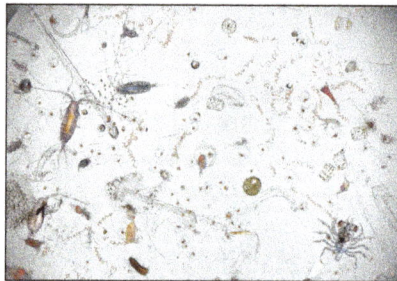

Under a magnifier, a splash of seawater teems with life.

Microscopic organisms live throughout the biosphere. The mass of prokaryote microorganisms — which includes bacteria and archaea, but not the nucleated eukaryote microorganisms — may be as much as 0.8 trillion tons of carbon (of the total biosphere mass, estimated at between 1 and 4 trillion tons). Single-celled barophilic marine microbes have been found at a depth of 10,900 m (35,800 ft) in the Mariana Trench, the deepest spot in the Earth's oceans. Microorganisms live inside rocks 580 m (1,900 ft) below the sea floor under 2,590 m (8,500 ft) of ocean off the coast of the northwestern United States, as well as 2,400 m (7,900 ft; 1.5 mi) beneath the seabed off Japan. The greatest known temperature at which microbial life can exist is 122 °C (252 °F). In 2014, scientists confirmed the existence of microorganisms living 800 m (2,600 ft) below the ice of Antarctica. According to one researcher, "You can find microbes everywhere — they're extremely adaptable to conditions, and survive wherever they are."

Marine Viruses

Viruses are small infectious agents that do not have their own metabolism and can replicate only inside the living cells of other organisms. Viruses can infect all types of life forms, from animals and plants to microorganisms, including bacteria and archaea. The linear size of the average virus is about one one-hundredth that of the average bacterium. Most viruses cannot be seen with an optical microscope so electron microscopes are used instead.

Viruses are found wherever there is life and have probably existed since living cells first evolved. The origin of viruses is unclear because they do not form fossils, so molecular techniques have been used to compare the DNA or RNA of viruses and are a useful means of investigating how they arise.

Viruses are now recognised as ancient and as having origins that pre-date the divergence of life into the three domains. But the origins of viruses in the evolutionary history of life are unclear: some may have evolved from plasmids—pieces of DNA that can move between cells—while others may have evolved from bacteria. In evolution, viruses are an important means of horizontal gene transfer, which increases genetic diversity.

Bacteriophages (phages)

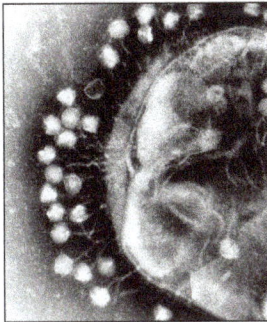

Multiple phages attached to a bacterial cell wall at 200,000x magnification.

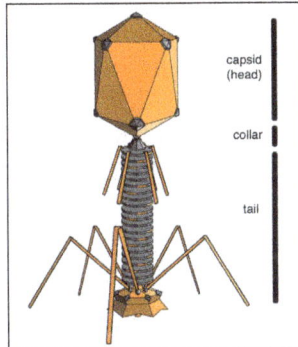

Diagram of a typical tailed phage.

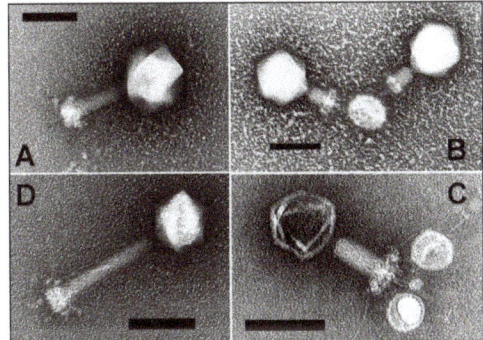

These are cyanophages, viruses that infect cyanobacteria (scale bars indicate 100 nm).

Opinions differ on whether viruses are a form of life or organic structures that interact with living organisms. They are considered by some to be a life form, because they carry genetic material, reproduce by creating multiple copies of themselves through self-assembly, and evolve through natural selection. However they lack key characteristics such as a cellular structure generally considered necessary to count as life. Because they possess some but not all such qualities, viruses have been described as replicators and as "organisms at the edge of life".

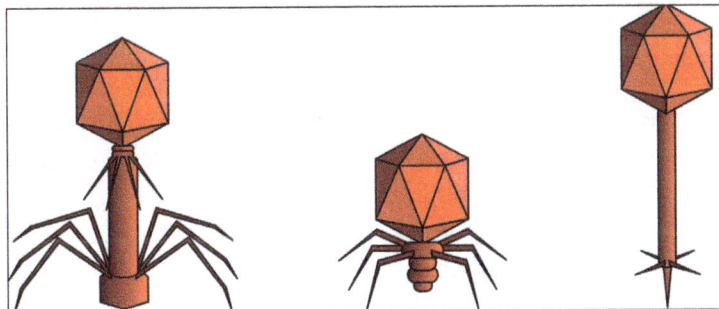

In terms of individual counts, tailed phage are the most abundant biological entities in the sea.

Bacteriophages, often just called *phages*, are viruses that parasite bacteria and archaea. Marine phages parasite marine bacteria and archaea, such as cyanobacteria. They are a common and diverse group of viruses and are the most abundant biological entity in marine environments, because their hosts, bacteria, are typically the numerically dominant cellular life in the sea. Generally there are about 1 million to 10 million viruses in each mL of seawater, or about ten times more double-stranded DNA viruses than there are cellular organisms, although estimates of viral abundance in seawater can vary over a wide range. Tailed bacteriophages appear to dominate marine ecosystems in number and diversity of organisms. Bacteriophages belonging to the families Corticoviridae, Inoviridae and Microviridae are also known to infect diverse marine bacteria.

Microorganisms make up about 70% of the marine biomass. It is estimated viruses kill 20% of this biomass each day and that there are 15 times as many viruses in the oceans as there are bacteria and archaea. Viruses are the main agents responsible for the rapid destruction of harmful algal blooms, which often kill other marine life. The number of viruses in the oceans decreases further offshore and deeper into the water, where there are fewer host organisms.

There are also archaean viruses which replicate within archaea: these are double-stranded DNA viruses with unusual and sometimes unique shapes. These viruses have been studied in most detail in the thermophilic archaea, particularly the orders Sulfolobales and Thermoproteales.

Viruses are an important natural means of transferring genes between different species, which increases genetic diversity and drives evolution. It is thought that viruses played a central role in the early evolution, before the diversification of bacteria, archaea and eukaryotes, at the time of the last universal common ancestor of life on Earth. Viruses are still one of the largest reservoirs of unexplored genetic diversity on Earth.

Marine Bacteria

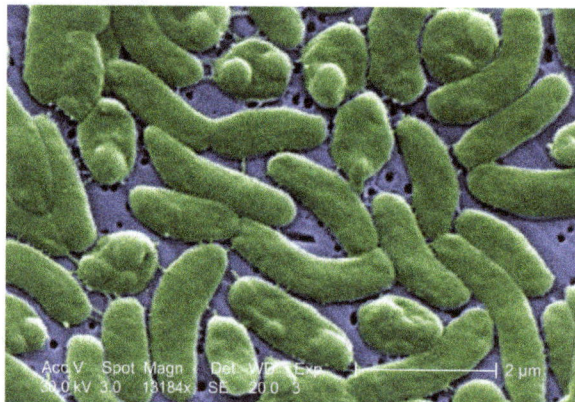

Vibrio vulnificus, a virulent bacterium found in estuaries and along coastal areas.

Bacteria constitute a large domain of prokaryotic microorganisms. Typically a few micrometres in length, bacteria have a number of shapes, ranging from spheres to rods and spirals. Bacteria were among the first life forms to appear on Earth, and are present in most of its habitats. Bacteria inhabit soil, water, acidic hot springs, radioactive waste, and the deep portions of Earth's crust. Bacteria also live in symbiotic and parasitic relationships with plants and animals.

Once regarded as plants constituting the class *Schizomycetes*, bacteria are now classified as pro-karyotes. Unlike cells of animals and other eukaryotes, bacterial cells do not contain a nucleus and rarely harbour membrane-bound organelles. Although the term *bacteria* traditionally included all prokaryotes, the scientific classification changed after the discovery in the 1990s that prokaryotes consist of two very different groups of organisms that evolved from an ancient common ancestor. These evolutionary domains are called *Bacteria* and *Archaea*.

The ancestors of modern bacteria were unicellular microorganisms that were the first forms of life to appear on Earth, about 4 billion years ago. For about 3 billion years, most organisms were microscopic, and bacteria and archaea were the dominant forms of life. Although bacterial fossils exist, such as stromatolites, their lack of distinctive morphology prevents them from being used to examine the history of bacterial evolution, or to date the time of origin of a particular bacterial species. However, gene sequences can be used to reconstruct the bacterial phylogeny, and these studies indicate that bacteria diverged first from the archaeal/eukaryotic lineage. Bacteria were also involved in the second great evolutionary divergence, that of the archaea and eukaryotes. Here, eukaryotes resulted from the entering of ancient bacteria into endosymbiotic associations with the ancestors of eukaryotic cells, which were themselves possibly related to the Archaea. This involved the engulfment by proto-eukaryotic cells of alphaproteobacterial symbionts to form either mitochondria or hydrogenosomes, which are still found in all known Eukarya. Later on, some eukaryotes that already contained mitochondria also engulfed cyanobacterial-like organisms. This led to the formation of chloroplasts in algae and plants. There are also some algae that originated from even later endosymbiotic events. Here, eukaryotes engulfed a eukaryotic algae that developed into a "second-generation" plastid. This is known as secondary endosymbiosis.

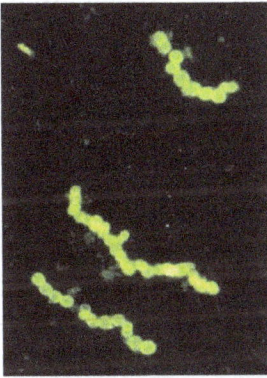

The marine Thiomargarita namibiensis, largest known bacterium.

Cyanobacteria blooms can contain lethal cyanotoxins.

The chloroplasts of glaucophytes have a peptidoglycan layer, evidence suggesting their endosymbiotic origin from cyanobacteria.

Bacteria can be beneficial. This Pompeii worm, an extremophile found only at hydrothermal vents, has a protective cover of bacteria.

The largest known bacterium, the marine *Thiomargarita namibiensis*, can be visible to the naked eye and sometimes attains 0.75 mm (750 μm).

Marine Archaea

The archaea constitute a domain and kingdom of single-celled microorganisms. These microbes are prokaryotes, meaning they have no cell nucleus or any other membrane-bound organelles in their cells.

Archaea were initially viewed as extremophiles living in harsh environments, such as the yellow archaea pictured here in a hot spring, but they have since been found in a much broader range of habitats.

Archaea were initially classified as bacteria, but this classification is outdated. Archaeal cells have unique properties separating them from the other two domains of life, Bacteria and Eukaryota. The Archaea are further divided into multiple recognized phyla. Classification is difficult because the majority have not been isolated in the laboratory and have only been detected by analysis of their nucleic acids in samples from their environment.

Archaea and bacteria are generally similar in size and shape, although a few archaea have very strange shapes, such as the flat and square-shaped cells of *Haloquadratum walsbyi*. Despite this morphological similarity to bacteria, archaea possess genes and several metabolic pathways that are more closely related to those of eukaryotes, notably the enzymes involved in transcription and translation. Other aspects of archaeal biochemistry are unique, such as their reliance on ether lipids in their cell membranes, such as archaeols. Archaea use more energy sources than eukaryotes: these range from organic compounds, such as sugars, to ammonia, metal ions or even hydrogen gas. Salt-tolerant archaea (the Haloarchaea) use sunlight as an energy source, and other species of archaea fix carbon; however, unlike plants and cyanobacteria, no known species of archaea does both. Archaea reproduce asexually by binary fission, fragmentation, or budding; unlike bacteria and eukaryotes, no known species forms spores.

Archaea are particularly numerous in the oceans, and the archaea in plankton may be one of the most abundant groups of organisms on the planet. Archaea are a major part of Earth's life and may play roles in both the carbon cycle and the nitrogen cycle.

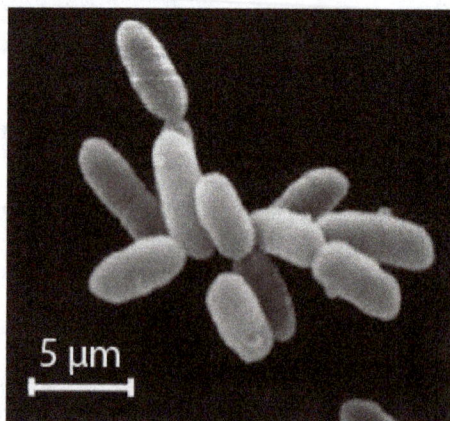

5 μm

Halobacteria, found in water near saturated with salt, are now recognised as archaea.

Flat, square-shaped cells of the
archaea Haloquadratum walsbyi.

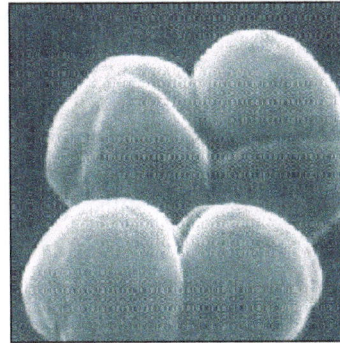

Methanosarcina barkeri, a marine archaea
that produces methane.

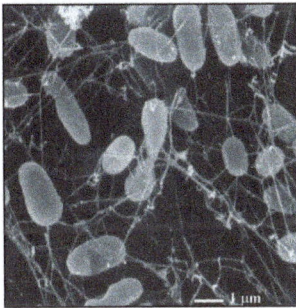

Thermophiles, such as Pyrolobus fumarii, survive
well over 100 °C.

Drawing of another marine thermophile,
Pyrococcus furiosus.

Marine Protists

All living organisms can be grouped as prokaryotes or eukaryotes. Life originated as single-celled prokaryotes (bacteria and archaea) and later evolved into more complex eukaryotes. Eukaryotes are the more developed life forms known as plants, animals, fungi and protists. In contrast to the cells of prokaryotes, the cells of eukaryotes are highly organised.

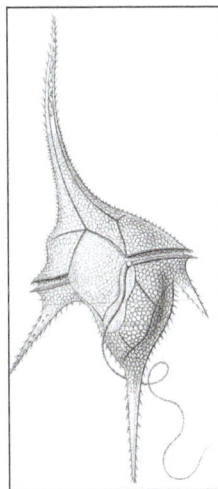

Dinoflagellate.

The term "protist" is came into use historically as a term of convenience for eukaryotes that cannot be classified as plants or as animals or as fungi. They are not a part of modern cladistics, because they are paraphyletic (lacking a common ancestor). However, protists can be put into three groups depending on whether their nutrition is animal-like or plant-like or fungus-like:

- Algae are the plant-like protists. They are autotrophs that make their own food without needing to consume other organisms, usually by using photosynthesis.

- Protozoa are the animal-like protists. They are heterotrophs that need to get their food by consuming other organisms.

- Slime moulds and water moulds are the fungus-like protists. Like protozoa they are heterotrophs, but they get their food from the remains of life forms that have broken down and decayed.

Microscopic Protists

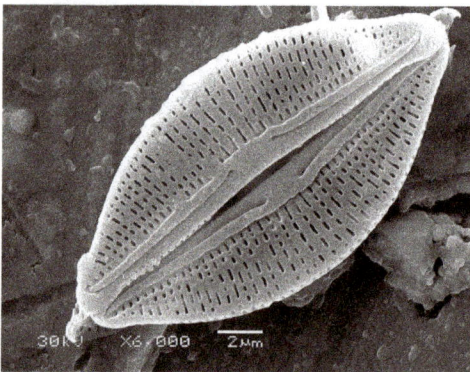
Diatoms have glass like cell walls made of silica and called frustules.

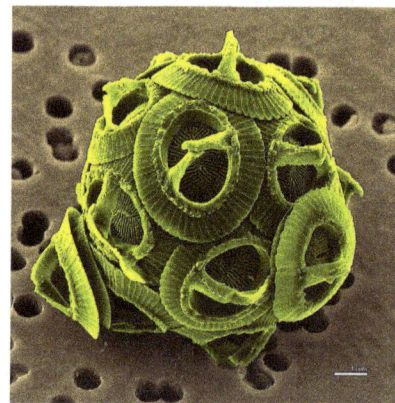
Single-celled alga, Gephyrocapsa oceanica.

Red algae, Cyanidium sp.

Protists include amoebae, ciliates, red algae, euglena, slime molds, and phytoplankton such as diatoms and dinoflagellates. They are a highly diverse group of organisms currently organised into 18 phyla, but they are not easy to classify. Studies have shown a high protist diversity exists in oceans, deep sea-vents and river sediments, which suggests a large number of eukaryotic microbial communities have yet to be discovered.

Plants, animals and fungi are usually multi-celled and are typically macroscopic. Most protists are single-celled and microscopic. But there are exceptions. Some single-celled marine protists are macroscopic. Some marine slime molds have unique life cycles that involve switching between unicellular, colonial, and multicellular forms. Other marine protist are neither single-celled nor microscopic, such as seaweed.

Macroscopic Protists

The unicellular bubble algae lives in tidal zones. It can have a 4 cm diameter.

The single-cell xenophyophore lives in abyssal zones. It has a giant shell up to 20 cm across.

In common usage the term protist is often used loosely to refer to unicellular eukaryotic microorganisms, even though technically some protists are multicellular and/or macroscopic.

Marine Microanimals

As juveniles, animals develop from microscopic stages, which can include spores, eggs and larvae. At least one microscopic animal group, the parasitic cnidarian Myxozoa, is unicellular in its adult form, and includes marine species. Other adult marine microanimals are multicellular. Microscopic adult arthropods are more commonly found inland in freshwater, but there are marine species as well. Microscopic adult marine crustaceans include some copepods, cladocera and water bears. Some marine nematodes and rotifers are also too small to be seen with the naked eye, as are many loricifera, including the recently discovered anaerobic species that spend their lives in an anoxic environment. Copepods contribute more to the secondary productivity and carbon sink of the world oceans than any other group of organisms.

Marine Micro-Animals

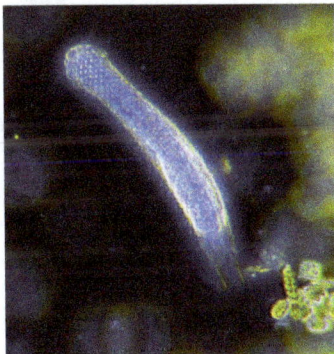

Darkfield photo of a gastrotrich, which form a phylum of worm-like animals living mostly between particles of sediment.

Armoured *Pliciloricus enigmaticus* of the phylum Loricifera live in the spaces between marine gravel.

Fungi

Lichen on a rock in a marine splash zone. Lichens are mutualistic associations between a fungus and an alga or cyanobacterium.

Over 1500 species of fungi are known from marine environments. These are parasitic on marine algae or animals, or are saprobes feeding on dead organic matter from algae, corals, protozoan cysts, sea grasses, wood and other substrata. Spores of many species have special appendages which facilitate attachment to the substratum. They can also be found in sea foam. A diverse range of unusual secondary metabolites is produced by marine fungi.

Mycoplankton are saprotropic members of the plankton communities of marine and freshwater ecosystems. They are composed of filamentous free-living fungi and yeasts that are associated with planktonic particles or phytoplankton. Similar to bacterioplankton, these aquatic fungi play a significant role in heterotrophic mineralization and nutrient cycling. Mycoplankton can be up to 20 mm in diameter and over 50 mm in length.

In a typical milliliter of seawater, there are approximately 10^3 to 10^4 fungal cells. This number is greater in coastal ecosystems and estuaries due to nutritional runoff from terrestrial communities. The greatest diversity and number of species of mycoplankton is found in surface waters (< 1000 m), and the vertical profile depends on the abundance of phytoplankton. Furthermore, this difference in distribution may vary between seasons due to nutrient availability. Marine fungi survive in a constant oxygen deficient environment, and therefore depend on oxygen diffusion by turbulence and oxygen generated by photosynthetic organisms.

Marine fungi can be classified as:

- Lower fungi: Adapted to marine habitats (zoosporic fungi, including mastigomycetes: oomycetes and chytridiomycetes).

- Higher fungi: Filamentous, modified to planktonic lifestyle (hyphomycetes, ascomycetes, basidiomycetes). Most mycoplankton species are higher fungi.

Lichens are mutualistic associations between a fungus, usually an ascomycete, and an alga or a cyanobacterium. Several lichens are found in marine environments. Many more occur in the splash zone, where they occupy different vertical zones depending on how tolerant they are to submersion.

According to fossil records, fungi date back to the late Proterozoic era 900-570 million years ago. Fossil marine lichens 600 million years old have been discovered in China. It has been hypothesized that mycoplankton evolved from terrestrial fungi, likely in the Paleozoic era (390 million years ago).

Invertebrate Animals

Dickinsonia may be the earliest animal. They appear in the fossil record 571 million to 541 million years ago.

The earliest animals were marine invertebrates, that is, vertebrates came later. Animals are multicellular eukaryotes, and are distinguished from plants, algae, and fungi by lacking cell walls. Marine invertebrates are animals that inhabit a marine environment apart from the vertebrate members of the chordate phylum; invertebrates lack a vertebral column. Some have evolved a shell or a hard exoskeleton.

The earliest animal fossils may belong to the genus *Dickinsonia*, 571 million to 541 million years ago. Individual *Dickinsonia* typically resemble a bilaterally symmetrical ribbed oval. They kept growing until they were covered with sediment or otherwise killed, and spent most of their lives with their bodies firmly anchored to the sediment. Their taxonomic affinities are presently unknown, but their mode of growth is consistent with a bilaterian affinity.

Apart from *Dickinsonia*, the earliest widely accepted animal fossils are the rather modern-looking cnidarians (the group that includes jellyfish, sea anemones and *Hydra*), possibly from around 580 Ma The Ediacara biota, which flourished for the last 40 million years before the start of the Cambrian, were the first animals more than a very few centimetres long. Like *Dickinsonia*, many were flat with a "quilted" appearance, and seemed so strange that there was a proposal to classify them as a separate kingdom, Vendozoa. Others, however, have been interpreted as early molluscs (*Kimberella*), echinoderms (*Arkarua*), and arthropods (*Spriggina, Parvancorina*). There is still debate about the classification of these specimens, mainly because the diagnostic features which allow taxonomists to classify more recent organisms, such as similarities to living organisms, are generally absent in the Ediacarans. However, there seems little doubt that *Kimberella* was at least a triploblastic bilaterian animal, in other words, an animal significantly more complex than the cnidarians.

Small shelly fauna are a very mixed collection of fossils found between the Late Ediacaran and Middle Cambrian periods. The earliest, Cloudina, shows signs of successful defense against predation and may indicate the start of an evolutionary arms race. Some tiny Early Cambrian shells almost certainly belonged to molluscs, while the owners of some "armor plates," Halkieria and Microdictyon, were eventually identified when more complete specimens were found in Cambrian lagerstätten that preserved soft-bodied animals.

Body Plans and Phyla

Kimberella, an early mollusc important for understanding the Cambrian explosion. Invertebrates are grouped into different phyla (body plans).

Opabinia, an extinct stem group arthropod appeared in the Middle Cambrian.

Invertebrates are grouped into different phyla. Informally phyla can be thought of as a way of grouping organisms according to their body plan. A body plan refers to a blueprint which describes the shape or morphology of an organism, such as its symmetry, segmentation and the disposition of its appendages. The idea of body plans originated with vertebrates, which were grouped into one phylum. But the vertebrate body plan is only one of many, and invertebrates consist of many phyla or body plans. The history of the discovery of body plans can be seen as a movement from a worldview centred on vertebrates, to seeing the vertebrates as one body plan among many. Among the pioneering zoologists, Linnaeus identified two body plans outside the vertebrates; Cuvier identified three; and Haeckel had four, as well as the Protista with eight more, for a total of twelve. For comparison, the number of phyla recognised by modern zoologists has risen to 35.

Historically body plans were thought of as having evolved rapidly during the Cambrian explosion, but a more nuanced understanding of animal evolution suggests a gradual development of body plans throughout the early Palaeozoic and beyond. More generally a phylum can be defined in two ways: as described above, as a group of organisms with a certain degree of morphological or developmental similarity (the phenetic definition), or a group of organisms with a certain degree of evolutionary relatedness (the phylogenetic definition).

In the 1970s there was already a debate about whether the emergence of the modern phyla was "explosive" or gradual but hidden by the shortage of Precambrian animal fossils. A re-analysis of fossils from the Burgess Shale lagerstätte increased interest in the issue when it revealed animals, such as *Opabinia*, which did not fit into any known phylum. At the time these were interpreted as evidence that the modern phyla had evolved very rapidly in the Cambrian explosion and that the Burgess Shale's "weird wonders" showed that the Early Cambrian was a uniquely experimental period of animal evolution. Later discoveries of similar animals and the development of new theoretical approaches led to the conclusion that many of the "weird wonders" were evolutionary "aunts" or "cousins" of modern groups—for example that *Opabinia* was a member of the lobopods, a group which includes the ancestors of the arthropods, and that it may have been closely related to the modern tardigrades. Nevertheless, there is still much debate about whether the Cambrian explosion was really explosive and, if so, how and why it happened and why it appears unique in the history of animals.

Arthropods total about 1,113,000 described extant species, molluscs about 85,000 and chordates about 52,000.

Marine Sponges

Sponges have no nervous, digestive or circulatory system.

Sponges are animals of the phylum Porifera. They are multicellular organisms that have bodies full of pores and channels allowing water to circulate through them, consisting of jelly-like mesohyl sandwiched between two thin layers of cells. They have unspecialized cells that can transform into other types and that often migrate between the main cell layers and the mesohyl in the process. Sponges do not have nervous, digestive or circulatory systems. Instead, most rely on maintaining a constant water flow through their bodies to obtain food and oxygen and to remove wastes.

Sponges are similar to other animals in that they are multicellular, heterotrophic, lack cell walls and produce sperm cells. Unlike other animals, they lack true tissues and organs, and have no body symmetry. The shapes of their bodies are adapted for maximal efficiency of water flow through the central cavity, where it deposits nutrients, and leaves through a hole called the osculum. Many sponges have internal skeletons of spongin and/or spicules of calcium carbonate or silicon dioxide. All sponges are sessile aquatic animals. Although there are freshwater species, the great majority are marine (salt water) species, ranging from tidal zones to depths exceeding 8,800 m (5.5 mi).

While most of the approximately 5,000–10,000 known species feed on bacteria and other food particles in the water, some host photosynthesizing micro-organisms as endosymbionts and these alliances often produce more food and oxygen than they consume. A few species of sponge that live in food-poor environments have become carnivores that prey mainly on small crustaceans.

Sponge biodiversity.

Branching vase sponge.

Stove-pipe sponge.

Linnaeus mistakenly identified sponges as plants in the order Algae. For a long time thereafter sponges were assigned to a separate subkingdom, Parazoa (meaning beside the animals). They are now classified as a paraphyletic phylum from which the higher animals have evolved.

Marine Cnidarians

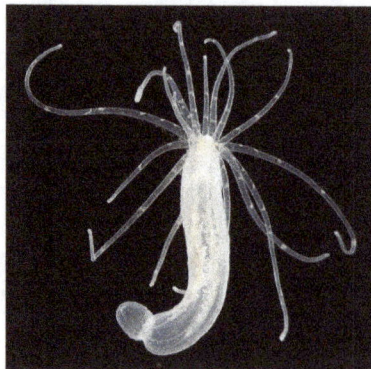
Cnidarians are the simplest animals with cells organised into tissues.
Yet the starlet sea anemone contains the same genes as those that form the vertebrate head.

Cnidarians are distinguished by the presence of stinging cells, specialized cells that they use mainly for capturing prey. Cnidarians include corals, sea anemones, jellyfish and hydrozoans. They form a phylum containing over 10,000 species of animals found exclusively in aquatic (mainly marine) environments. Their bodies consist of mesoglea, a non-living jelly-like substance, sandwiched between two layers of epithelium that are mostly one cell thick. They have two basic body forms: swimming medusae and sessile polyps, both of which are radially symmetrical with mouths surrounded by tentacles that bear cnidocytes. Both forms have a single orifice and body cavity that are used for digestion and respiration.

Sea anemones are common in tidepools.

Their tentacles sting and paralyse small fish.

Fossil cnidarians have been found in rocks formed about 580 million years ago. Fossils of cnidarians that do not build mineralized structures are rare. Scientists currently think cnidarians, ctenophores and bilaterians are more closely related to calcareous sponges than these are to other sponges, and that anthozoans are the evolutionary "aunts" or "sisters" of other cnidarians, and the most closely related to bilaterians.

Cnidarians are the simplest animals in which the cells are organised into tissues. The starlet sea anemone is used as a model organism in research. It is easy to care for in the laboratory and a protocol has been developed which can yield large numbers of embryos on a daily basis. There is a remarkable degree of similarity in the gene sequence conservation and complexity between the sea anemone and vertebrates. In particular, genes concerned in the formation of the head in vertebrates are also present in the anemone.

Close up of polyps on the surface of a coral,
waving their tentacles.

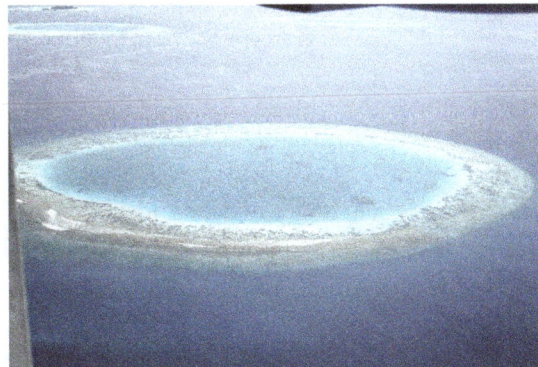

If an island sinks below the sea, coral growth can keep
up with rising water and form an atoll.

Porpita porpita consists of a colony of hydroids.

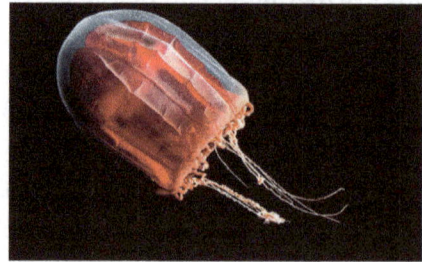
The mantle of the red paper lantern jellyfish crumples and expands like a paper lantern.

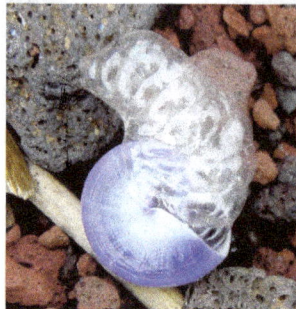
Turritopsis dohrnii achieves biological immortality by transferring its cells back to childhood.

Marine Worms

Worms (Old English for *serpent*) form a number of phylums. Their body plan typically involves long cylindrical tube-like bodies and no limbs. Marine worms vary in size from microscopic to over 1 metre (3.3 ft) in length for some marine polychaete worms (bristle worms) and up to 58 metres (190 ft) for the marine nemertean worm (bootlace worm). Some marine worms occupy a small variety of parasitic niches, living inside the bodies of other animals, while others live more freely in the marine environment or by burrowing underground.

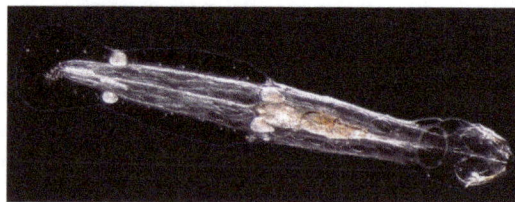
Many marine worms are related only distantly, so they form a number of different phyla. The worm shown is an arrow worm, found worldwide as a predatory component of plankton.

Different groups of marine worms are related only distantly, so they are found in several different phyla such as the Annelida (segmented worms), Chaetognatha (arrow worms), Hemichordata, and Phoronida (horseshoe worms). Many of these worms have specialized tentacles used for exchanging oxygen and carbon dioxide and also may be used for reproduction. Some marine worms are tube worms, such as the giant tube worm which lives in waters near underwater volcanoes and can withstand temperatures up to 90 degrees Celsius.

Platyhelminthes (flatworms) form another worm phylum which includes a class Cestoda of parasitic tapeworms. The marine tapeworm *Polygonoporus giganticus*, found in the gut of sperm whales, can grow to over 30 m (100 ft).

Nematodes (roundworms) constitute a further worm phylum with tubular digestive systems and an opening at both ends. Over 25,000 nematode species have been described, of which more than half are parasitic. It has been estimated another million remain undescribed. They are ubiquitous in marine, freshwater and terrestrial environments, where they often outnumber other animals in both individual and species counts. They are found in every part of the earth's lithosphere, from the top of mountains to the bottom of oceanic trenches. By count they represent 90% of all animals on the ocean floor. Their numerical dominance, often exceeding a million individuals per square meter and accounting for about 80% of all individual animals on earth, their diversity of life cycles, and their presence at various trophic levels point at an important role in many ecosystems.

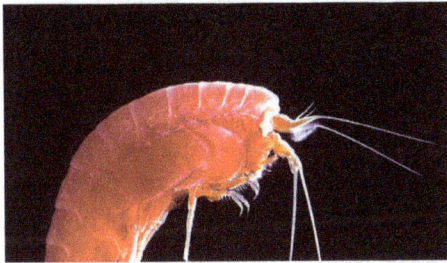

Giant tube worms cluster around hydrothermal vents.

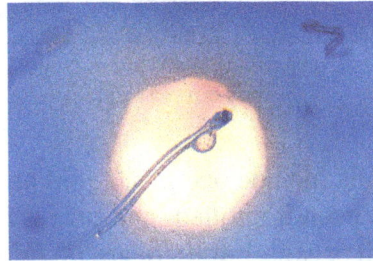

Bloodworms are typically found on the bottom of shallow marine waters.

Marine Molluscs

Bigfin reef squid displaying vivid iridescence at night. Cephalopods are the most neurologically advanced invertebrates.

Molluscs form a phylum with about 85,000 extant recognized species. They are the largest marine phylum in terms of species count, comprising about 23% of all the named marine organisms. Molluscs have more varied forms than other invertebrate phylums. They are highly diverse, not just in size and in anatomical structure, but also in behaviour and in habitat. The majority of species still live in the oceans, from the seashores to the abyssal zone, but some form a significant part of the freshwater fauna and the terrestrial ecosystems.

The mollusc phylum is divided into 9 or 10 taxonomic classes, two of which are extinct. These classes include gastropods, bivalves and cephalopods, as well as other lesser-known but distinctive classes. Gastropods with protective shells are referred to as snails, whereas gastropods without protective shells are referred to as slugs. Gastropods are by far the most numerous molluscs in terms of classified species, accounting for 80% of the total. Bivalves include clams, oysters, cockles, mussels, scallops, and numerous other families. There are about 8,000 marine bivalves species (including brackish water and estuarine species), and about 1,200 freshwater species.

Gastropods and Bivalves

Marine gastropods are sea snails or sea slugs. This nudibranch is a sea slug.

Molluscs usually have eyes. Bordering the edge of the mantle of a scallop, a bivalve mollusc, can be over 100 simple eyes.

Cephalopods include octopus, squid and cuttlefish. About 800 living species of marine cephalopods have been identified, and an estimated 11,000 extinct taxa have been described. They are found in all oceans, but there are no fully freshwater cephalopods.

Cephalopods

Molluscs have such diverse shapes that many textbooks base their descriptions of molluscan anatomy on a generalized or *hypothetical ancestral mollusc*. This generalized mollusc is unsegmented and bilaterally symmetrical with an underside consisting of a single muscular foot. Beyond that it has three further key features. Firstly, it has a muscular cloak called a mantle covering its viscera and containing a significant cavity used for breathing and excretion. A shell secreted by the mantle covers the upper surface. Secondly (apart from bivalves) it has a rasping tongue called a radula used for feeding. Thirdly, it has a nervous system including a complex digestive system using microscopic, muscle-powered hairs called cilia to exude mucus. The generalized mollusc has two paired nerve cords (three in bivalves). The brain, in species that have one, encircles the esophagus.

Most molluscs have eyes and all have sensors detecting chemicals, vibrations, and touch. The simplest type of molluscan reproductive system relies on external fertilization, but more complex variations occur. All produce eggs, from which may emerge trochophore larvae, more complex veliger larvae, or miniature adults. The depiction is rather similar to modern monoplacophorans, and some suggest it may resemble very early molluscs.

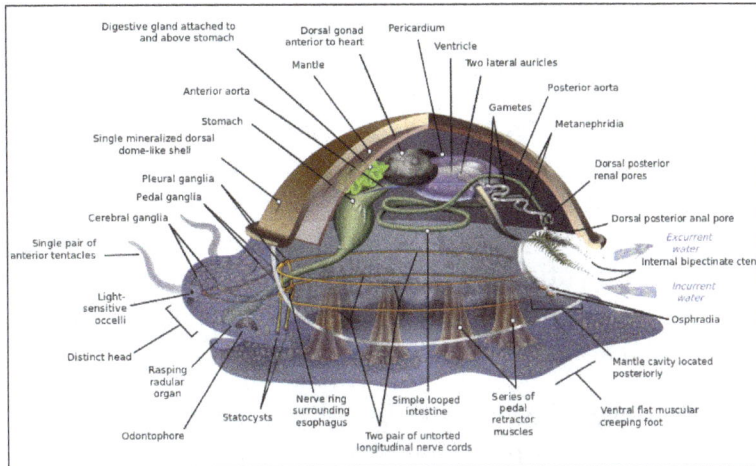

Generalized or hypothetical ancestral mollusc.

Good evidence exists for the appearance of marine gastropods, cephalopods and bivalves in the Cambrian period 541 to 485.4 million years ago. However, the evolutionary history both of molluscs' emergence from the ancestral Lophotrochozoa and of their diversification into the well-known living and fossil forms are still subjects of vigorous debate among scientists.

Marine Arthropods

Segments and tagmata of an arthropod. The head and thorax are fused in some arthropods, such as crabs and lobsters.

Pneumodesmus newmani is the first known invertebrate to colonise land. It lived in the Early Devonian.

Arthropods have an exoskeleton (external skeleton), a segmented body, and jointed appendages (paired appendages). They form a phylum which includes insects, arachnids, myriapods, and crustaceans. Arthropods are characterized by their jointed limbs and cuticle made of chitin, often mineralised with calcium carbonate. The arthropod body plan consists of segments, each with a pair of appendages. The rigid cuticle inhibits growth, so arthropods replace it periodically by moulting. Their versatility has enabled them to become the most species-rich members of all ecological guilds in most environments.

Extant marine arthropods range in size from the microscopic crustacean *Stygotantulus* to the Japanese spider crab. Arthropods' primary internal cavity is a hemocoel, which accommodates their internal organs, and through which their haemolymph - analogue of blood - circulates; they have open circulatory systems. Like their exteriors, the internal organs of arthropods are generally built of repeated segments. Their nervous system is "ladder-like", with paired ventral nerve cords running through all segments and forming paired ganglia in each segment. Their heads are formed by fusion of varying numbers of segments, and their brains are formed by fusion of the ganglia of these segments and encircle the esophagus. The respiratory and excretory systems of arthropods vary, depending as much on their environment as on the subphylum to which they belong.

Arthropod vision relies on various combinations of compound eyes and pigment-pit ocelli: in most species the ocelli can only detect the direction from which light is coming, and the compound eyes are the main source of information. Arthropods also have a wide range of chemical and mechanical sensors, mostly based on modifications of the many setae (bristles) that project through their cuticles. Arthropod methods of reproduction are diverse: terrestrial species use some form of internal fertilization while marine species lay eggs using either internal or external fertilization. Arthropod hatchlings vary from miniature adults to grubs that lack jointed limbs and eventually undergo a total metamorphosis to produce the adult form.

Fossils and Living Fossils

Fossil trilobite. Trilobites first appeared about 521 Ma. They were highly successful and were found everywhere in the ocean for 270 Ma.

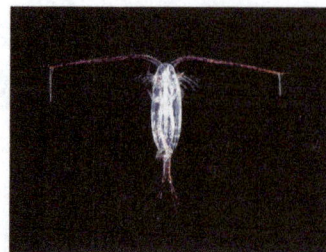

Horseshoe crabs are living fossils, essentially unchanged for 450 Ma.

The largest known arthropod, the sea scorpion *Jaekelopterus rhenaniae*, has been found in estuarine strata from about 390 Ma. It was up to 2.5 m (8.2 ft) long.

The *Anomalocaris* ("abnormal shrimp") was one of the first apex predators and first appeared about 515 Ma.

Crustaceans

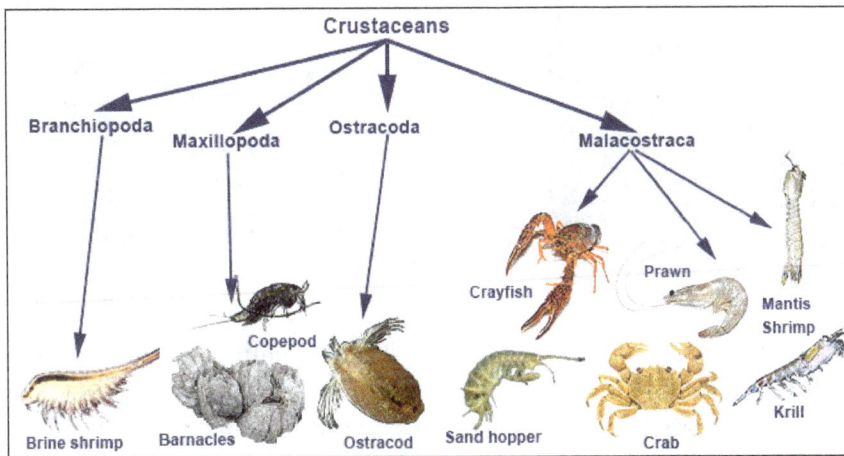

The evolutionary ancestry of arthropods dates back to the Cambrian period. The group is generally regarded as monophyletic, and many analyses support the placement of arthropods with cyclo-neuralians (or their constituent clades) in a superphylum Ecdysozoa. Overall however, the basal relationships of animals are not yet well resolved. Likewise, the relationships between various arthropod groups are still actively debated.

Echinoderms

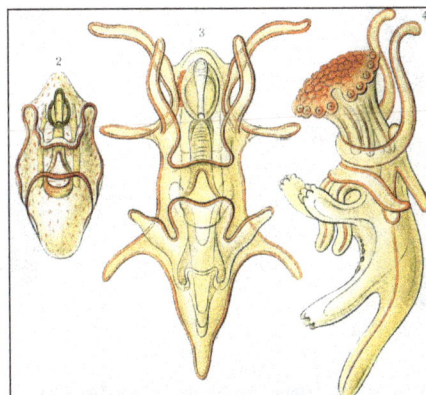

Starfish larvae are bilaterally symmetric, whereas the adults have fivefold symmetry.

Echinoderms is a phylum which contains only marine invertebrates. The adults are recognizable by their radial symmetry (usually five-point) and include starfish, sea urchins, sand dollars, and sea cucumbers, as well as the sea lilies. Echinoderms are found at every ocean depth, from the intertidal zone to the abyssal zone. The phylum contains about 7000 living species, making it the second-largest grouping of deuterostomes (a superphylum), after the chordates (which include the vertebrates, such as birds, fishes, mammals, and reptiles).

Echinoderms are unique among animals in having bilateral symmetry at the larval stage, but five-fold symmetry (pentamerism, a special type of radial symmetry) as adults.

Echinoderm literally means "spiny skin", as this water melon sea urchin illustrates.

The ochre sea star was the first keystone predator to be studied. They limit mussels which can overwhelm intertidal communities.

The echinoderms are important both biologically and geologically. Biologically, there are few other groupings so abundant in the biotic desert of the deep sea, as well as shallower oceans. Most echinoderms are able to regenerate tissue, organs, limbs, and reproduce asexually; in some cases, they can undergo complete regeneration from a single limb. Geologically, the value of echinoderms is in their ossified skeletons, which are major contributors to many limestone formations, and can provide valuable clues as to the geological environment. They were the most used species in regenerative research in the 19th and 20th centuries.

Colorful sea lilies in shallow waters.

Sea cucumbers filter feed on plankton and suspended solids.

It is held by some scientists that the radiation of echinoderms was responsible for the Mesozoic Marine Revolution. Aside from the hard-to-classify *Arkarua* (a Precambrian animal with echinoderm-like pentamerous radial symmetry), the first definitive members of the phylum appeared near the start of the Cambrian.

Marine Chordates

The chordate phylum has three subphylums, one of which is the vertebrates. The other two subphylums are marine invertebrates: the tunicates (salps and sea squirts) and the cephalochordates (such as lancelets). Invertebrate chordates are close relatives to vertebrates. In particular, there has been discussion about how closely some extinct marine species, such as Pikaiidae, Palaeospondylus, Zhongxiniscus and Vetulicolia, might relate ancestrally to vertebrates. Invertebrate chordates are close relatives of vertebrates.

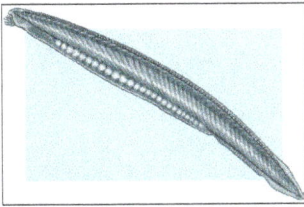

The lancelet, a small translucent fish-like cephalochordate, is the closest living invertebrate relative of the vertebrates.

Tunicates, like these fluorescent-colored sea squirts, may provide clues to vertebrate and therefore human ancestry.

Other Phyla

Other phyla containing marine invertebrates include tardigrades and comb jellies, as well as extinct phyla such as lobopodians. Lobopodians are an extinct group of worm-like taxa with stubby legs that resemble modern velvet worms. They date from the Lower Cambrian.

Some palaeontologists think Lobopodia represents a basal grade which lead to an arthropod body plan.

Tardigrades are a phylum of eight-legged, segmented micro-animals able to survive in extreme conditions.

Vertebrate Animals

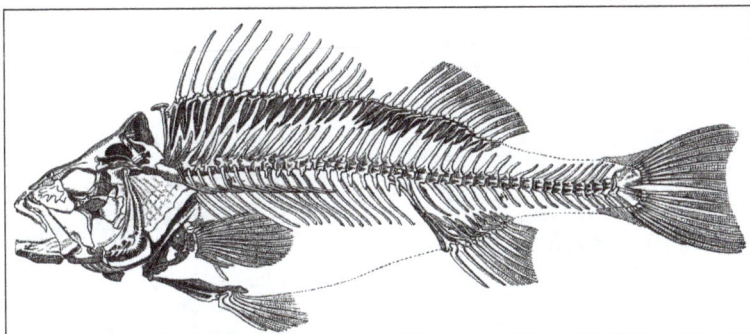

Ray-finned fish.

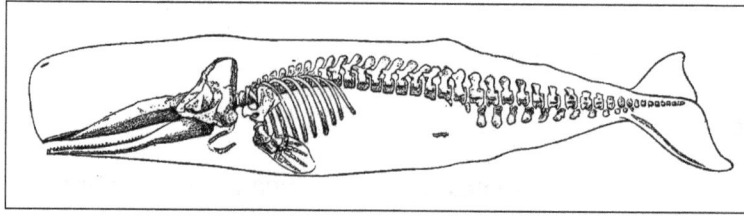

Marine tetrapod (sperm whale)
Skeletal structures showing the vertebral column running from the head to the tail.

Vertebrates are a subphylum of chordates. They are chordates that have a vertebral column (backbone). The vertebral column provides the central support structure for an internal skeleton. The internal skeleton gives shape, support, and protection to the body and can provide a means of anchoring fins or limbs to the body. The vertebral column also serves to house and protect the spinal cord that lies within the column.

Marine vertebrates can be divided into marine fish and marine tetrapods.

Marine Fish

Fish fall into two main groups: fish with bony internal skeletons and fish with cartilaginous internal skeletons. Fish anatomy and physiology generally includes a two-chambered heart, eyes adapted to seeing underwater, and a skin protected by scales and mucous. They typically breathe by extracting oxygen from water through gills. Fish use fins to propel and stabilise themselves in the water. Over 33,000 species of fish have been described as of 2017, of which about 20,000 are marine fish.

Jawless Fish

Hagfish form a class of about 20 species of eel-shaped, slime-producing marine fish. They are the only known living animals that have a skull but no vertebral column. Lampreys form a superclass containing 38 known extant species of jawless fish. The adult lamprey is characterized by a toothed, funnel-like sucking mouth. Although they are well known for boring into the flesh of other fish to suck their blood, only 18 species of lampreys are actually parasitic. Together hagfish and lampreys are the sister group to vertebrates. Living hagfish remain similar to hagfish from around 300 million years ago. The lampreys are a very ancient lineage of vertebrates, though their exact relationship to hagfishes and jawed vertebrates is still a matter of dispute. Molecular analysis since 1992 has suggested that hagfish are most closely related to lampreys, and so also are vertebrates in a monophyletic sense. Others consider them a sister group of vertebrates in the common taxon of craniata.

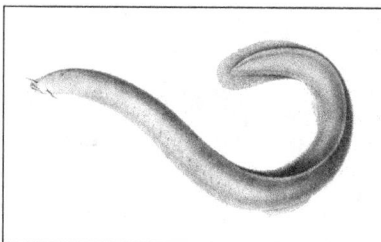

Hagfish are the only known living animals with a skull but no vertebral column.

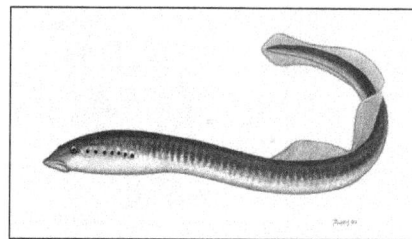

Lampreys are often parasitic and have a toothed, funnel-like sucking mouth .

Pteraspidomorphi is an extinct class of early jawless fish ancestral to jawed vertebrates. The few characteristics they share with the latter are now considered as primitive for all vertebrates.

Cartilaginous Fish

Cartilaginous fish, such as sharks and rays, have jaws and skeletons made of cartilage rather than bone. Megalodon is an extinct species of shark that lived about 28 to 1.5 Ma. It looked much like a stocky version of the great white shark, but was much larger with fossil lengths reaching 20.3 metres (67 ft). Found in all oceans it was one of the largest and most powerful predators in vertebrate history, and probably had a profound impact on marine life. The Greenland shark has the longest known lifespan of all vertebrates, about 400 years. Some sharks such as the great white are partially warm blooded and give live birth.

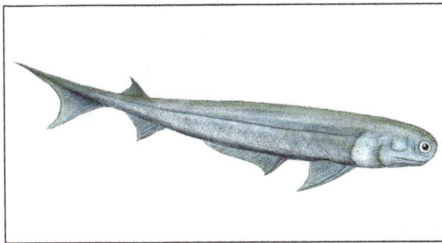

Cartilaginous fishes may have evolved from spiny sharks .

Stingray.

The largest extant fish, the whale shark, is now a vulnerable species.

Bony Fish

Bony fish have jaws and skeletons made of bone rather than cartilage. About 90% of the world's fish species are bony fish. Bony fish also have hard, bony plates called operculum which help them respire and protect their gills, and they often possess a swim bladder which they use for better control of their buoyancy.

Bony fish can be further divided into those with lobe fins and those with ray fins. Lobe fins have the form of fleshy lobes supported by bony stalks which extend from the body. Lobe fins evolved into the legs of the first tetrapod land vertebrates, so by extension an early ancestor of humans was a lobe-finned fish. Apart from the coelacanths and the lungfishes, lobe-finned fishes are now extinct. The rest of the modern fish have ray fins. These are made of webs of skin supported by bony or horny spines (rays) which can be erected to control the fin stiffness.

Ray-finned fish (Prussian carp).

Lobe-finned fish (coelacanth).

Sailfish.

Marine Tetrapods

A tetrapod is a vertebrate with limbs (feet). Tetrapods evolved from ancient lobe-finned fishes about 400 million years ago during the Devonian Period when their earliest ancestors emerged from the sea and adapted to living on land. This change from a body plan for breathing and navigating in gravity-neutral water to a body plan with mechanisms enabling the animal to breath in air without dehydrating and move on land is one of the most profound evolutionary changes known. Tetrapods can be divided into four classes: amphibians, reptiles, birds and mammals.

Tiktaalik, an extinct lobe-finned fish, developed limb-like fins that could take it onto land.

Marine tetrapods are tetrapods that returned from land back to the sea again. The first returns to the ocean may have occurred as early as the Carboniferous Period whereas other returns occurred as recently as the Cenozoic, as in cetaceans, pinnipeds, and several modern amphibians.

Amphibians

live part of their life in water and part on land. They mostly require fresh water to reproduce. A few inhabit brackish water, but there are no true marine amphibians. There have been reports, however, of amphibians invading marine waters, such as a Black Sea invasion by the natural hybrid *Pelophylax esculentus* reported in 2010.

Reptiles

Reptiles do not have an aquatic larval stage, and in this way are unlike amphibians. Most reptiles are oviparous, although several species of squamates are viviparous, as were some extinct aquatic clades—the fetus develops within the mother, contained in a placenta rather than an eggshell. As amniotes, reptile eggs are surrounded by membranes for protection and transport, which adapt them to reproduction on dry land. Many of the viviparous species feed their fetuses through various forms of placenta analogous to those of mammals, with some providing initial care for their hatchlings.

Some reptiles are more closely related to birds than other reptiles, and many scientists prefer to make Reptilia a monophyletic group which includes the birds. Extant non-avian reptiles which inhabit or frequent the sea include sea turtles, sea snakes, terrapins, the marine iguana, and the saltwater crocodile. Currently, of the approximately 12,000 extant reptile species and sub-species, only about 100 of are classed as marine reptiles.

Except for some sea snakes, most extant marine reptiles are oviparous and need to return to land to lay their eggs. Apart from sea turtles, the species usually spend most of their lives on or near land rather than in the ocean. Sea snakes generally prefer shallow waters nearby land, around islands, especially waters that are somewhat sheltered, as well as near estuaries. Unlike land snakes, sea snakes have evolved flattened tails which help them swim.

Marine iguana.

Saltwater crocodile.

Marine snakes have flattened tails.

Some extinct marine reptiles, such as ichthyosaurs, evolved to be viviparous and had no requirement to return to land. Ichthyosaurs resembled dolphins. They first appeared about 245 million years ago and disappeared about 90 million years ago. The terrestrial ancestor of the ichthyosaur had no features already on its back or tail that might have helped along the evolutionary process. Yet the ichthyosaur developed a dorsal and tail fin which improved its ability to swim. The biologist Stephen Jay Gould said the ichthyosaur was his favourite example of convergent evolution. The earliest marine reptiles arose in the Permian. During the Mesozoic many groups of reptiles became adapted to life in the seas, including ichthyosaurs, plesiosaurs, mosasaurs, nothosaurs, placodonts, sea turtles, thalattosaurs and thalattosuchians. Marine reptiles were less numerous after mass extinction at the end of the Cretaceous.

Birds

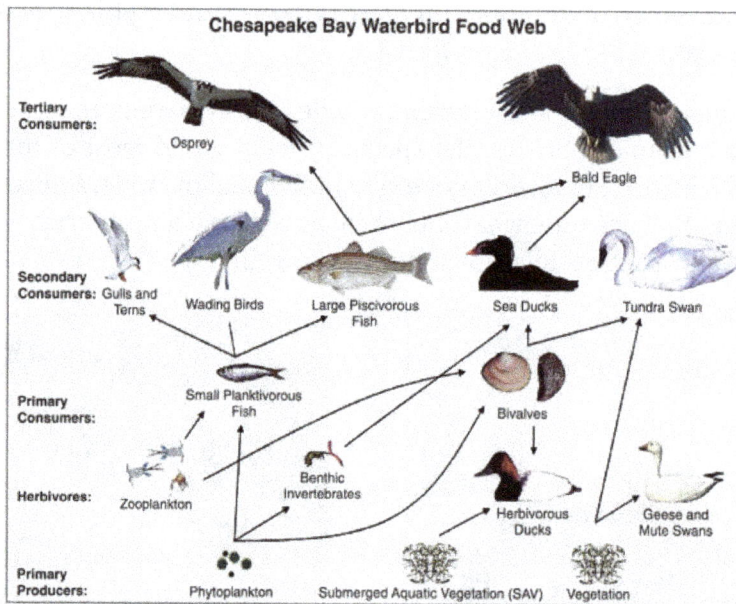

Waterbird food web in Chesapeake Bay.

Marine birds are adapted to life within the marine environment. They are often called *seabirds*. While marine birds vary greatly in lifestyle, behaviour and physiology, they often exhibit striking convergent evolution, as the same environmental problems and feeding niches have resulted in similar adaptations. Examples include albatross, penguins, gannets, and auks.

In general, marine birds live longer, breed later and have fewer young than terrestrial birds do, but they invest a great deal of time in their young. Most species nest in colonies, which can vary in size from a few dozen birds to millions. Many species are famous for undertaking long annual migrations, crossing the equator or circumnavigating the Earth in some cases. They feed both at the ocean's surface and below it, and even feed on each other. Marine birds can be highly pelagic, coastal, or in some cases spend a part of the year away from the sea entirely. Some marine birds plummet from heights, plunging through the water leaving vapour-like trails, similar to that of fighter planes. Gannets plunge into the water at up to 100 kilometres per hour (60 mph). They have air sacs under their skin in their face and chest which act like bubble-wrap, cushioning the impact with the water.

European herring gull attack herring schools from above.

Gentoo penguin swimming underwater.

The first marine birds evolved in the Cretaceous period, and modern marine bird families emerged in the Paleogene.

Mammals

Mammals are characterised by the presence of mammary glands which in females produce milk for feeding (nursing) their young. There are about 130 living and recently extinct marine mammal species such as seals, dolphins, whales, manatees, sea otters and polar bears. They do not represent a distinct taxon or systematic grouping, but are instead unified by their reliance on the marine environment for feeding. Both cetaceans and sirenians are fully aquatic and therefore are obligate water dwellers. Seals and sea-lions are semiaquatic; they spend the majority of their time in the water, but need to return to land for important activities such as mating, breeding and molting. In contrast, both otters and the polar bear are much less adapted to aquatic living. Their diet varies considerably as well: some may eat zooplankton; others may eat fish, squid, shellfish, and sea-grass; and a few may eat other mammals.

In a process of convergent evolution, marine mammals, especially cetaceans redeveloped their body plan to parallel the streamlined fusiform body plan of pelagic fish. Front legs became flippers and back legs disappeared, a dorsal fin reappeared and the tail morphed into a powerful horizontal fluke. This body plan is an adaptation to being an active predator in a high drag environment. A parallel convergence occurred with the now extinct marine reptile ichthyosaur.

Dugong grazing on seagrass.

Polar bear.

Primary Producers

Primary producers are the autotroph organisms that make their own food instead of eating other organisms. This means primary producers become the starting point in the food chain for heterotroph organisms that do eat other organisms. Some marine primary producers are specialised bacteria and archaea which are chemotrophs, making their own food by gathering around hydrothermal vents and cold seeps and using chemosynthesis. However most marine primary production comes from organisms which use photosynthesis on the carbon dioxide dissolved in the water. This process uses energy from sunlight to convert water and carbon dioxide into sugars that can be used both as a source of chemical energy and of organic molecules that are used in the structural components of cells. Marine primary producers are important because they underpin almost all marine animal life by generating most of the oxygen and food that provide other organisms with the chemical energy they need to exist.

The principal marine primary producers are cyanobacteria, algae and marine plants. The oxygen released as a by-product of photosynthesis is needed by nearly all living things to carry out cellular respiration. In addition, primary producers are influential in the global carbon and water cycles. They stabilize coastal areas and can provide habitats for marine animals. The term division has been traditionally used instead of phylum when discussing primary producers, although the International Code of Nomenclature for algae, fungi, and plants now accepts the terms as equivalent.

Cyanobacteria

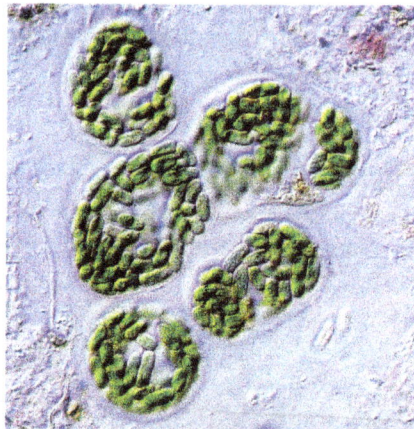

Cyanobacteria from a microbial mat. Cyanobacteria were the first organisms to release oxygen via photosynthesis.

Cyanobacteria are a phylum (division) of bacteria which range from unicellular to filamentous and include colonial species. They are found almost everywhere on earth: in damp soil, in both freshwater and marine environments, and even on Antarctic rocks. In particular, some species occur as drifting cells floating in the ocean, and as such were amongst the first of the phytoplankton.

The first primary producers that used photosynthesis were oceanic cyanobacteria about 2.3 billion years ago. The release of molecular oxygen by cyanobacteria as a by-product of photosynthesis induced global changes in the Earth's environment. Because oxygen was toxic to most life on Earth at the time, this led to the near-extinction of oxygen-intolerant organisms, a dramatic change which redirected the evolution of the major animal and plant species.

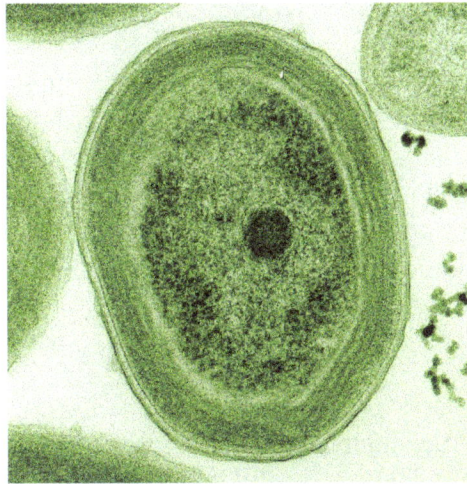

Prochlorococcus marinus.

The tiny marine cyanobacterium *Prochlorococcus*, discovered in 1986, forms today part of the base of the ocean food chain and accounts for more than half the photosynthesis of the open ocean and an estimated 20% of the oxygen in the Earth's atmosphere. It is possibly the most plentiful genus on Earth: a single millilitre of surface seawater may contain 100,000 cells or more.

Originally, biologists thought cyanobacteria was algae, and referred to it as "blue-green algae". The more recent view is that cyanobacteria is a bacteria, and hence is not even in the same Kingdom as algae. Most authorities exclude all prokaryotes, and hence cyanobacteria from the definition of algae.

Algae

Algae is an informal term for a widespread and diverse group of photosynthetic eukaryotic organisms which are not necessarily closely related and are thus polyphyletic. Algae can be divided into six groups:

- Green algae, an informal group containing about 8,000 recognized species. Many species live most of their lives as single cells or are filamentous, while others form colonies made up from long chains of cells, or are highly differentiated macroscopic seaweeds.

- Red algae, a (disputed) phylum containing about 7,000 recognized species, mostly multicellular and including many notable seaweeds.

- Brown algae, a class containing about 2,000 recognized species, mostly multicellular and including many seaweeds, including kelp.

- Diatoms.

- Dinoflagellates.

- Euglenophytes, a phylum of unicellular flagellates with only a few marine members.

Unlike higher plants, algae lack roots, stems, or leaves. They can be classified by size as *microalgae* or *macroalgae*.

Microalgae

Microalgae are the microscopic types of algae, not visible to the naked eye. They are mostly unicellular species which exist as individuals or in chains or groups, though some are multicellular. Microalgae are important components of the marine protists discussed above, as well as the phytoplankton discussed below. They are very diverse. It has been estimated there are 200,000-800,000 species of which about 50,000 species have been described. Depending on the species, their sizes range from a few micrometers (μm) to a few hundred micrometers. They are specially adapted to an environment dominated by viscous forces.

Macroalgae

Macroalgae are the larger, multicellular and more visible types of algae, commonly called seaweeds. Seaweeds usually grow in shallow coastal waters where they are anchored to the seafloor by a holdfast. Seaweed that becomes adrift can wash up on beaches. Kelp is a large brown seaweed that forms large underwater forests covering about 25% of the world coastlines. They are among the most productive and dynamic ecosystems on Earth. Some *Sargassum* seaweeds are planktonic (free-floating). Like microalgae, macroalgae (seaweeds) are technically marine protists since they are not true plants.

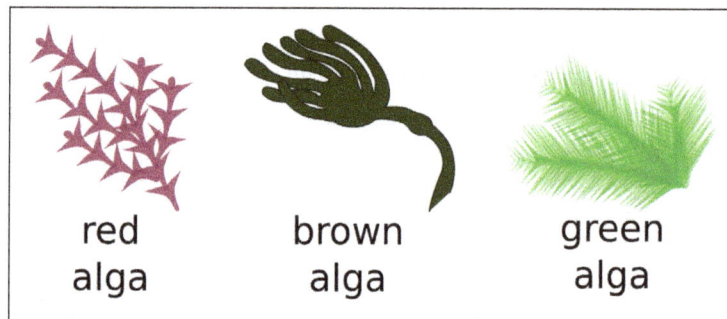

red
alga

brown
alga

green
alga

A seaweed is a macroscopic form of red or brown or green algae.

Marine Plants

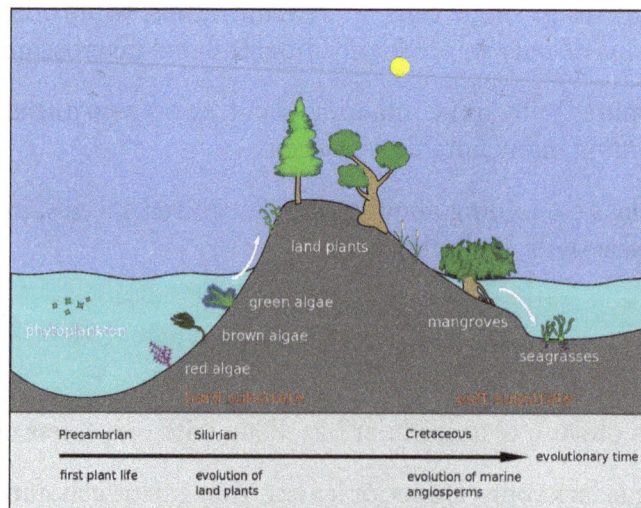

Evolution of mangroves and seagrasses.

Back in the Silurian, some phytoplankton evolved into red, brown and green algae. These algae then invaded the land and started evolving into the land plants we know today. Later, in the Cretaceous, some of these land plants returned to the sea as mangroves and seagrasses.

Marine plants can be found in intertidal zones and shallow waters, such as seagrasses like eelgrass and turtle grass, *Thalassia*. These plants have adapted to the high salinity of the ocean environment. Plant life can also flourish in the brackish waters of estuaries, where mangroves or cordgrass or beach grass beach grass might grow.

Mangroves.

Seagrass meadow.

The total world area of mangrove forests was estimated in 2010 as 134,257 square kilometres (51,837 sq mi) (based on satellite data). The total world area of seagrass meadows is more difficult to determine, but was conservatively estimated in 2003 as 177,000 square kilometres (68,000 sq mi).

Mangroves and seagrasses provide important nursery habitats for marine life, acting as hiding and foraging places for larval and juvenile forms of larger fish and invertebrates.

Plankton and Trophic Interactions

Plankton are drifting or floating organisms that cannot swim against a current, and include organisms from all the domains of life: bacteria, archaea, algae, protozoa and animal.

Plankton are a diverse group of organisms that live in the water column of large bodies of water but cannot swim against a current. As a result, they wander or drift with the currents. Plankton are defined by their ecological niche, not by any phylogenetic or taxonomic classification. They are a crucial source of food for many marine animals, from forage fish to whales. Plankton can be divided into a plant-like component and an animal component.

Phytoplankton

Phytoplankton are the plant-like components of the plankton community. They are autotrophic (self-feeding), meaning they generate their own food and do not need to consume other organisms.

Phytoplankton consist mainly of microscopic photosynthetic eukaryotes which inhabit the upper sunlit layer in all oceans. They need sunlight so they can photosynthesize. Most phytoplankton are single-celled algae, but other phytoplankton are bacteria and some are protists. Phytoplankton can be categorised into cyanobacteria (also called blue-green algae/bacteria), various types of algae (red, green, brown, and yellow-green), diatoms, dinoflagellates, euglenoids, coccolithophorids, cryptomonads, chrysophytes, chlorophytes, prasinophytes, and silicoflagellates. They form the base of the primary production that drives the ocean food web, and account for half of the current global primary production, more than the terrestrial forests.

Zooplankton

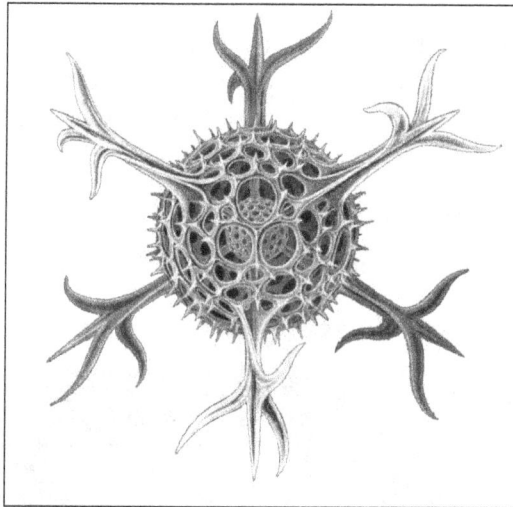

Radiolarian protist drawn by Ernst Haeckel.

Zooplankton are the animal component of the planktonic community. They are heterotrophic (other-feeding), meaning they cannot produce their own own food and must consume instead other plants or animals as food. In particular, this means they eat phytoplankton.

Zooplankton are generally larger than phytoplankton, mostly still microscopic but some can be seen with the naked eye. Many protozoans (single-celled eukaryotes) are zooplankton, including dinoflagellates, zooflagellates, foraminiferans, and radiolarians. Some of these, such as dinoflagellates, can also be classified as phytoplankton; the distinction between plants and animals often breaks down in very small organisms. Other zooplankton include cnidarians, ctenophores, chaetognaths, molluscs, arthropods, urochordates, and annelids such as polychaetes.

Microzooplankton: Major Grazers of the Plankton

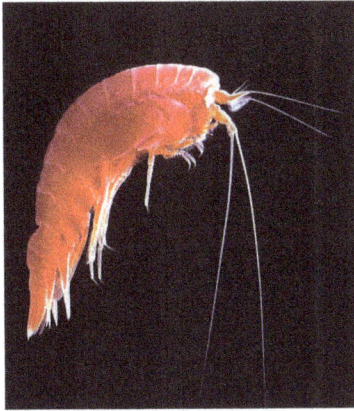

Marine amphipod.

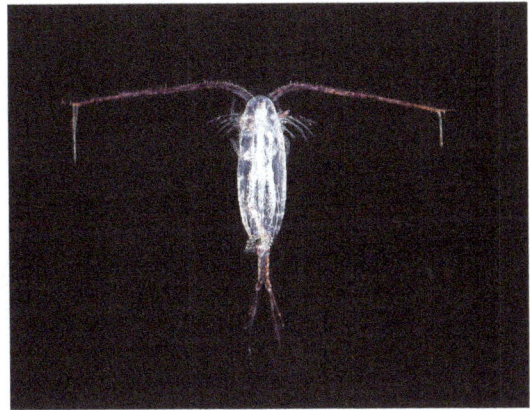

Copepod.

Larger zooplankton can be predatory and eat smaller zooplankton.

Macrozooplankton

Tomopteris, a planktonic segmented worm with unusual yellow bioluminescence.

Many marine animals begin life as zooplankton in the form of eggs or larvae, before they develop into adults.

Spawn, Larvae and Juveniles

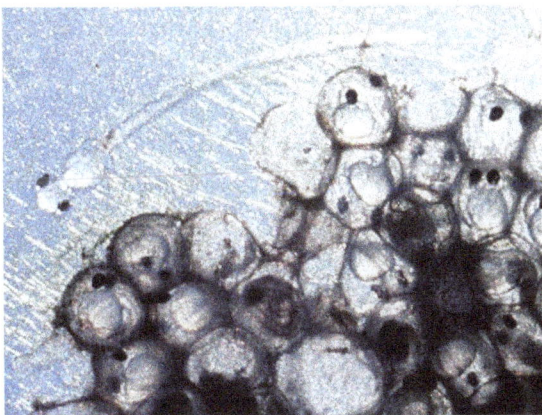

Transparent herring eggs with yolk and eyes visible and one larva hatched.

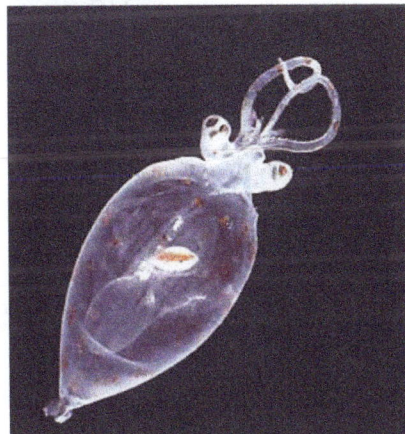

Juvenile planktonic squid.

Marine Food Web

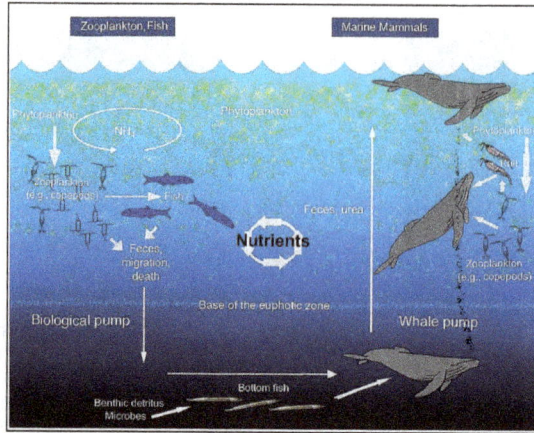

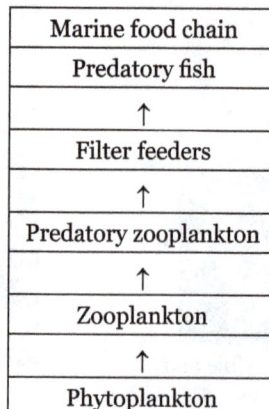

Whale pump nutrient cycle.

In 2010 researchers found whales carry nutrients from the depths of the ocean back to the surface using a process they called the whale pump. Whales feed at deeper levels in the ocean where krill is found, but return regularly to the surface to breathe. There whales defecate a liquid rich in nitrogen and iron. Instead of sinking, the liquid stays at the surface where phytoplankton consume it. In the Gulf of Maine the whale pump provides more nitrogen than the rivers.

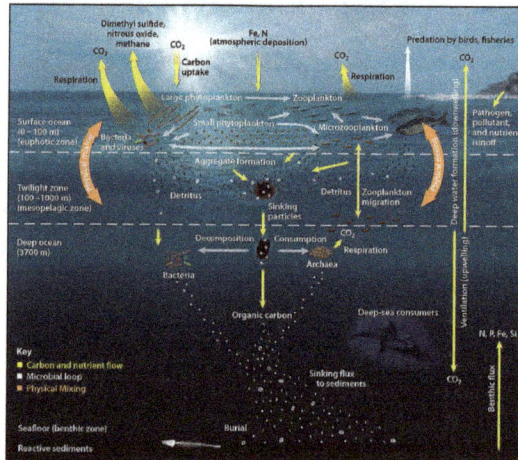

Marine food web.

Marine food chain
Predatory fish
↑
Filter feeders
↑
Predatory zooplankton
↑
Zooplankton
↑
Phytoplankton

If phytoplankton dies before it is eaten, it descends through the euphotic zone and settles into the depths of sea. In this way, phytoplankton sequester about 2 billion tons of carbon dioxide into the ocean each year, causing the ocean to become a sink of carbon dioxide holding about 90% of all sequestered carbon.

Zooplankton make up most of the marine animal biomass, and as primary consumers are the crucial link between primary producers (mainly phytoplankton) and the rest of the marine food web (secondary consumers).

Biogeochemical Cycles

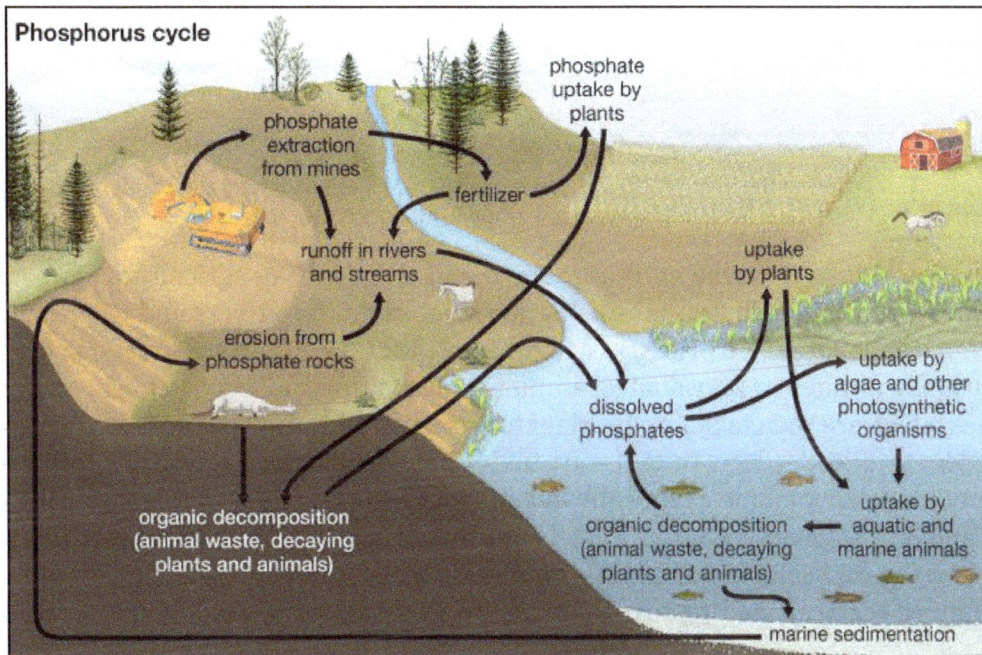

In 2000 a team of microbiologists led by Edward DeLong made a crucial discovery in the understanding of the marine carbon and energy cycles. They discovered a gene in several species of bacteria responsible for production of the protein rhodopsin, previously unheard of in the domain Bacteria. These proteins found in the cell membranes are capable of converting light energy to biochemical energy due to a change in configuration of the rhodopsin molecule as sunlight strikes it, causing the pumping of a proton from inside out and a subsequent inflow that generates the energy.

Anthropogenic Impacts

"It is almost as though we use our military to fight the animals in the ocean. We are gradually winning this war to exterminate them."

Ocean acidification is the increasing acidification of the oceans, caused by the uptake of carbon dioxide from the atmosphere. The rise in atmospheric carbon dioxide due to the burning of fossil fuels leads to more carbon dioxide dissolving in the ocean. When carbon dioxide dissolves in water it forms hydrogen and carbonate ions. This in turn increases the acidity of the ocean and makes survival increasingly harder for shellfish and other marine organisms that depend on calcium carbonate to form their shells. Increasing acidity is also thought to have a range of potentially harmful consequences for marine organisms, such as depressing metabolic rates and immune responses in some organisms, and causing coral bleaching. Ocean acidification has been compared to anthropogenic climate change and called the "evil twin of global warming" and "the other CO_2 problem".

Fishing down the foodweb: Overfishing of high trophic fish like tuna can result in them being replaced by low trophic organisms, like jellyfish.

Marine pollution results from the entry into the ocean of industrial, agricultural, and residential wastes. Pathways for this pollution include agricultural runoff into rivers and wind-blown debris and dust. Nutrient pollution is a primary cause of eutrophication of surface waters, in which excess nutrients, usually nitrates or phosphates, stimulate algae growth. Toxic chemicals can adhere to tiny particles which are then taken up by plankton and benthic animals, most of which are either deposit feeders or filter feeders. In this way, toxins are concentrated upward within ocean food chains. Many particles combine chemically in a manner which depletes oxygen, causing estuaries to become anoxic. Pesticides and toxic metals are similarly incorporated into marine food webs, harming the biological health of marine life. Many animal feeds have a high fish meal or fish hydrolysate content. In this way, marine toxins are transferred back to farmed land animals, and then to humans.

Phytoplankton concentrations have increased over the last century in coastal waters, and more recently have declined in the open ocean. Increases in nutrient runoff from land may explain the increases in coastal phytoplankton, while warming surface temperatures in the open ocean may have strengthened stratification in the water column, reducing the flow of nutrients from the deep that open ocean phytoplankton find useful.

Estimates suggest something like 9 million tonnes of plastic is added to the ocean every year. It is thought this plastic will need 450 years or more to biograde. Once in the ocean, plastics are shredded by marine amphipods into microplastics. There are now beaches where 15 percent of the sand are grains of microplastic. In the oceans themselves, microplastics float in surface waters amongst the plankton, where they are ingested by plankton eaters.

Overfishing is occurring in one third of world fish stocks, according to a 2018 report by the Food and Agriculture Organization of the United Nations. In addition, industry observers believe illegal, unreported and unregulated fishing occurs in most fisheries, and accounts for up to 30% of total catches in some important fisheries. In a phenomenon called fishing down the foodweb, the mean trophic level of world fisheries has declined because of overfishing high trophic level fish.

Habitat loss is occurring in seagrass meadows, mangrove forests, coral reefs and kelp forests, which are in global decline due to human disturbances. Seagrasses have lost 30,000 km² (12,000 sq mi) during recent decades, one fifth of the world's mangrove forests have been lost since 1980, and another fifth of coral reefs have been lost. The most pressing threat to kelp forests may be the overfishing of coastal ecosystems, which by removing higher trophic levels facilitates their shift to depauperate urchin barrens.

Shifting baselines arise in research on marine ecosystems because changes must be measured against some previous reference point (baseline), which in turn may represent significant changes from an even earlier state of the ecosystem. For example, radically depleted fisheries have been evaluated by researchers who used the state of the fishery at the start of their careers as the baseline, rather than the fishery in its unexploited or untouched state. Areas that swarmed with a particular species hundreds of years ago may have experienced long term decline, but it is the level a few decades previously that is used as the reference point for current populations. In this way large declines in ecosystems or species over long periods of time were, and are, masked. There is a loss of perception of change that occurs when each generation redefines what is natural or untouched.

Biodiversity and Extinction Events

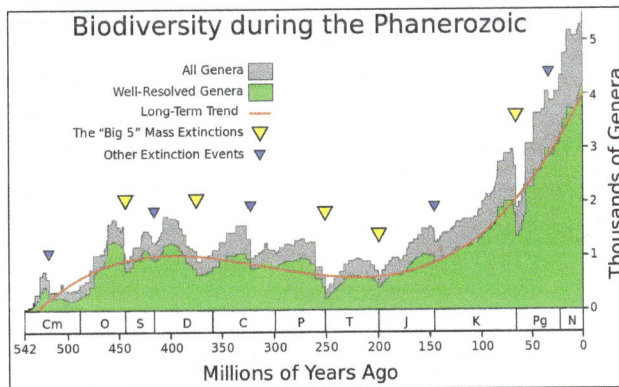

Apparent marine fossil diversity during the Phanerozoic.

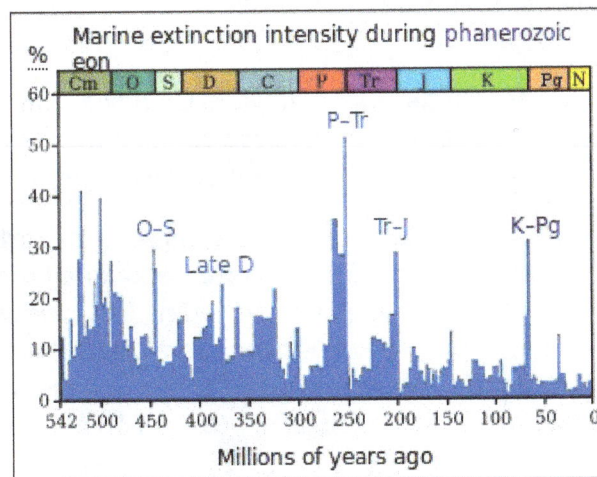

Millions of years ago: Apparent extinction intensity, i.e. the fraction of genera going extinct at any given time as reconstructed from the fossil record (excluding the current Holocene extinction event).

Biodiversity is the result of over three billion years of evolution. Until approximately 600 million years ago, all life consisted of archaea, bacteria, protozoans and similar single-celled organisms. The history of biodiversity during the Phanerozoic (the last 540 million years), starts with rapid growth during the Cambrian explosion: a period during which nearly every phylum of multicellular organisms first appeared. Over the next 400 million years or so, invertebrate diversity showed little overall trend and vertebrate diversity shows an overall exponential trend.

However, more than 99 percent of all species that ever lived on Earth, amounting to over five billion species, are estimated to be extinct. These extinctions occur at an uneven rate. The dramatic rise in diversity has been marked by periodic, massive losses of diversity classified as mass extinction events. Mass extinction events occur when life undergoes precipitous global declines. Most diversity and biomass on earth is found among the microorganisms, which are difficult to measure. Recorded extinction events are therefore based on the more easily observed changes in the diversity and abundance of larger multicellular organisms, rather than the total diversity and abundance of life. Marine fossils are mostly used to measure extinction rates because of their superior fossil record and stratigraphic range compared to land organisms.

Based on the fossil record, the background rate of extinctions on Earth is about two to five taxonomic families of marine animals every million years. The Great Oxygenation Event was perhaps the first major extinction event. Since the Cambrian explosion five further major mass extinctions have significantly exceeded the background extinction rate. The worst was the Permian-Triassic extinction event, 251 million years ago. Vertebrates took 30 million years to recover from this event. In addition to these major mass extinctions there are numerous minor ones, as well as the current ongoing mass-extinction caused by human activity, the Holocene extinction sometimes called the "sixth extinction".

Coral Reef

A coral reef is an underwater ecosystem characterized by reef-building corals. Reefs are formed of colonies of coral polyps held together by calcium carbonate. Most coral reefs are built from stony corals, whose polyps cluster in groups.

Coral belongs to the class *Anthozoa* in the animal phylum *Cnidaria*, which includes sea anemones and jellyfish. Unlike sea anemones, corals secrete hard carbonate exoskeletons that support and protect the coral. Most reefs grow best in warm, shallow, clear, sunny and agitated water.

Often called "rainforests of the sea", shallow coral reefs form some of Earth's most diverse ecosystems. They occupy less than 0.1% of the world's ocean area, about half the area of France, yet they provide a home for at least 25% of all marine species, including fish, mollusks, worms, crustaceans, echinoderms, sponges, tunicates and other cnidarians. Coral reefs flourish in ocean waters that provide few nutrients. They are most commonly found at shallow depths in tropical waters, but deep water and cold water coral reefs exist on smaller scales in other areas.

Coral reefs deliver ecosystem services for tourism, fisheries and shoreline protection. The annual global economic value of coral reefs is estimated between US$30–375 billion and 9.9 trillion USD.

Coral reefs are fragile, partly because they are sensitive to water conditions. They are under threat from excess nutrients (nitrogen and phosphorus), rising temperatures, oceanic acidification, over-fishing (e.g., from blast fishing, cyanide fishing, spearfishing on scuba), sunscreen use, and harm-ful land-use practices, including runoff and seeps (e.g., from injection wells and cesspools).

Formation

Most coral reefs were formed after the last glacial period when melting ice caused sea level to rise and flood continental shelves. Most coral reefs are less than 10,000 years old. As communities established themselves, the reefs grew upwards, pacing rising sea levels. Reefs that rose too slowly could become drowned, without sufficient light. Coral reefs are found in the deep sea away from continental shelves, around oceanic islands and atolls. The majority of these islands are volcanic in origin. Others have tectonic origins where plate movements lifted the deep ocean floor.

In *The Structure and Distribution of Coral Reefs*, Charles Darwin set out his theory of the forma-tion of atoll reefs, an idea he conceived during the voyage of the *Beagle*. He theorized that uplift and subsidence of the Earth's crust under the oceans formed the atolls. Darwin set out a sequence of three stages in atoll formation. A fringing reef forms around an extinct volcanic island as the island and ocean floor subsides. As the subsidence continues, the fringing reef becomes a barrier reef and ultimately an atoll reef.

Darwin's theory starts with a volcanic island which becomes extinct.

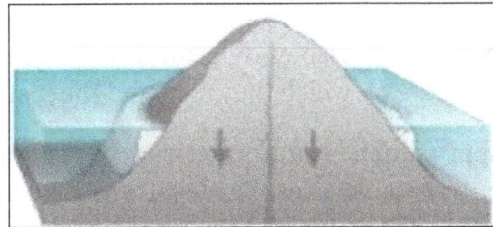

As the island and ocean floor subside, coral growth builds a fringing reef, often including a shallow lagoon between the land and the main reef.

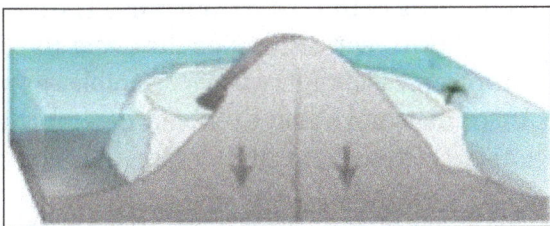

As the subsidence continues, the fringing reef becomes a larger barrier reef further from the shore with a bigger and deeper lagoon inside.

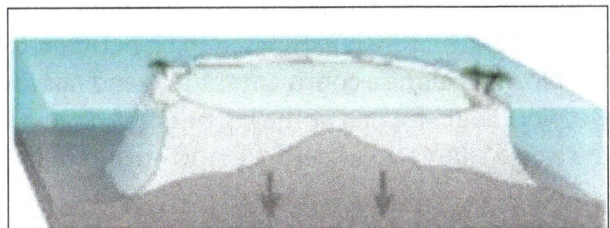

Ultimately, the island sinks below the sea, and the barri-er reef becomes an atoll enclosing an open lagoon.

Darwin predicted that underneath each lagoon would be a bedrock base, the remains of the orig-inal volcano. Subsequent research supported this hypothesis. Darwin's theory followed from his understanding that coral polyps thrive in the tropics where the water is agitated, but can only live within a limited depth range, starting just below low tide. Where the level of the underlying earth allows, the corals grow around the coast to form fringing reefs, and can eventually grow to become a barrier reef.

Coral Reef.

Where the bottom is rising, fringing reefs can grow around the coast, but coral raised above sea level dies. If the land subsides slowly, the fringing reefs keep pace by growing upwards on a base of older, dead coral, forming a barrier reef enclosing a lagoon between the reef and the land. A barrier reef can encircle an island, and once the island sinks below sea level a roughly circular atoll of growing coral continues to keep up with the sea level, forming a central lagoon. Barrier reefs and atolls do not usually form complete circles, but are broken in places by storms. Like sea level rise, a rapidly subsiding bottom can overwhelm coral growth, killing the coral and the reef, due to what is called *coral drowning*. Corals that rely on zooxanthellae can die when the water becomes too deep for their symbionts to adequately photosynthesize, due to decreased light exposure.

The two main variables determining the geomorphology, or shape, of coral reefs are the nature of the substrate on which they rest, and the history of the change in sea level relative to that substrate.

The approximately 20,000-year-old Great Barrier Reef offers an example of how coral reefs formed on continental shelves. Sea level was then 120 m (390 ft) lower than in the 21st century. As sea level rose, the water and the corals encroached on what had been hills of the Australian coastal plain. By 13,000 years ago, sea level had risen to 60 m (200 ft) lower than at present, and many hills of the coastal plains had become continental islands. As sea level rise continued, water topped most of the continental islands. The corals could then overgrow the hills, forming cays and reefs. Sea level on the Great Barrier Reef has not changed significantly in the last 6,000 years. The age of living reef structure is estimated to be between 6,000 and 8,000 years. Although the Great Barrier Reef formed along a continental shelf, and not around a volcanic island, Darwin's principles apply. Development stopped at the barrier reef stage, since Australia is not about to submerge. It formed the world's largest barrier reef, 300–1,000 m (980–3,280 ft) from shore, stretching for 2,000 km (1,200 mi).

Healthy tropical coral reefs grow horizontally from 1 to 3 cm (0.39 to 1.18 in) per year, and grow vertically anywhere from 1 to 25 cm (0.39 to 9.84 in) per year; however, they grow only at depths shallower than 150 m (490 ft) because of their need for sunlight, and cannot grow above sea level.

Material

As the name implies, coral reefs are made up of coral skeletons from mostly intact coral colonies. As other chemical elements present in corals become incorporated into the calcium carbonate deposits, aragonite is formed. However, shell fragments and the remains of coralline algae such as the green-segmented genus *Halimeda* can add to the reef's ability to withstand damage from storms and other threats. Such mixtures are visible in structures such as Eniwetok Atoll.

Types

Since Darwin's identification of the three classical reef formations: the fringing reef around a volcanic island becoming a barrier reef and then an atoll: scientists have identified further reef types. While some sources find only three, Thomas and Goudie list four "principal large-scale coral reef types": the fringing reef, barrier reef, atoll and table reef: while Spalding *et al.* list five "main types": the fringing reef, barrier reef, atoll, "bank or platform reef" and patch reef.

Fringing Reef

Fringing reef.

Fringing reef at Eilat at the southern tip of Israel.

A fringing reef, also called a shore reef, is directly attached to a shore, or borders it with an intervening narrow, shallow channel or lagoon. It is the most common reef type. Fringing reefs follow coastlines and can extend for many kilometres. They are usually less than 100 metres wide, but some are hundreds of metres wide. Fringing reefs are initially formed on the shore at the low water level and expand seawards as they grow in size. The final width depends on where the sea bed begins to drop steeply. The surface of the fringe reef generally remains at the same height: just below the waterline. In older fringing reefs, whose outer regions pushed far out into the sea, the inner part is deepened by erosion and eventually forms a lagoon. Fringing reef lagoons can become over 100 metres wide and several metres deep. Like the fringing reef itself, they run parallel to the coast. The fringing reefs of the Red Sea are "some of the best developed in the world" and occur along all its shores except off sandy bays.

Barrier Reef

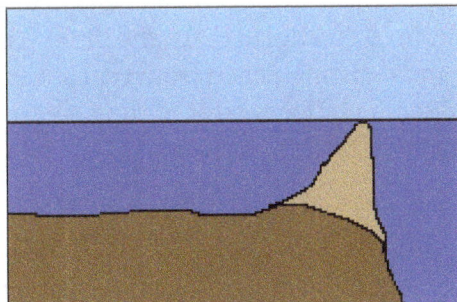

Barrier reef.

Barrier reefs are separated from a mainland or island shore by a deep channel or lagoon. They resemble the later stages of a fringing reef with its lagoon, but differ from the latter mainly in size and origin. Their lagoons can be several kilometres wide and 30 to 70 metres deep. Above all, the offshore outer reef edge formed in open water rather than next to a shoreline. Like an atoll, it is thought that these reefs are formed either as the seabed lowered or sea level rose. Formation takes considerably longer than for a fringing reef, thus barrier reefs are much rarer.

The best known and largest example of a barrier reef is the Australian Great Barrier Reef. Other major examples are the Belize Barrier Reef and the New Caledonian Barrier Reef. Barrier reefs are also found on the coasts of Providencia, Mayotte, the Gambier Islands, on the southeast coast of Kalimantan, on parts of the coast of Sulawesi, southeastern New Guinea and the south coast of the Louisiade Archipelago.

Platform Reef

Platform reef.

Platform reefs, variously called bank or table reefs, can form on the continental shelf, as well as in the open ocean, in fact anywhere where the seabed rises close enough to the surface of the ocean to enable the growth of zooxanthemic, reef-forming corals. Platform reefs are found in the southern Great Barrier Reef, the Swain and Capricorn Group on the continental shelf, about 100–200 km from the coast. Some platform reefs of the northern Mascarenes are several thousand kilometres from the mainland. Unlike fringing and barrier reefs which extend only seaward, platform reefs grow in all directions. They are variable in size, ranging from a few hundred metres to many kilometres across. Their usual shape is oval to elongated. Parts of these reefs can reach the surface and form sandbanks and small islands around which may form fringing reefs. A lagoon may form In the middle of a platform reef.

Platform reefs can be found within atolls. There they are called patch reefs and may reach only a few dozen metres in diameter. Where platform reefs form on an elongated structure, e. g. an old, eroded barrier reef, they can form a linear arrangement. This is the case, for example, on the east coast of the Red Sea near Jeddah. In old platform reefs, the inner part can be so heavily eroded that it forms a pseudo-atoll. These can be distinguished from real atolls only by detailed investigation, possibly including core drilling. Some platform reefs of the Laccadives are U-shaped, due to wind and water flow.

Atoll

Atolls or atoll reefs are a more or less circular or continuous barrier reef that extends all the way around a lagoon without a central island. They are usually formed from fringing reefs around volcanic islands. Over time, the island erodes away and sinks below sea level. Atolls may also be formed by the sinking of the seabed or rising of the sea level. A ring of reefs results, which enclose a lagoon. Atolls are numerous in the South Pacific, where they usually occur in mid-ocean, for example, in the Caroline Islands, the Cook Islands, French Polynesia, the Marshall Islands and Micronesia.

Formation of an atoll according to Charles Darwin.

Atolls are found in the Indian Ocean, for example, in the Maldives, the Chagos Islands, the Seychelles and around Cocos Island. The entire Maldives consist of 26 atolls.

Other Reef Types or Variants

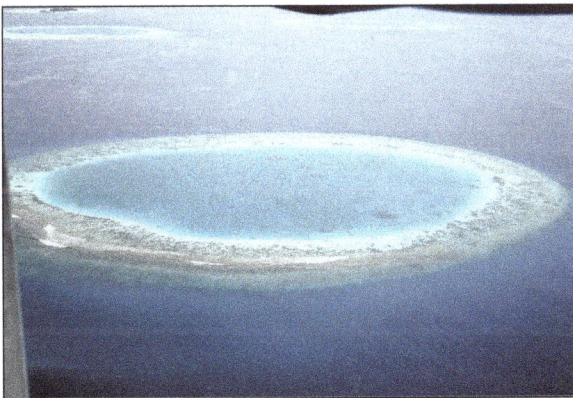

A small atoll in the Maldives.

Inhabited cay in the Maldives.

- Apron reef: Short reef resembling a fringing reef, but more sloped; extending out and downward from a point or peninsular shore. The initial stage of a fringing reef.

- Bank reef: Isolated, flat-topped reef larger than a patch reef and usually on mid-shelf regions and linear or semi-circular in shape; a type of platform reef.

- Patch reef: Common, isolated, comparatively small reef outcrop, usually within a lagoon or embayment, often circular and surrounded by sand or seagrass. Can be considered as a type of platform reef or as features of fringing reefs, atolls and barrier reefs. The patches may be surrounded by a ring of reduced seagrass cover refereed to as a grazing halo.

- Ribbon reef: Long, narrow, possibly winding reef, usually associated with an atoll lagoon. Also called a shelf-edge reef or sill reef.

- Habili: Reef specific to the Red Sea; does not reach near enough to the surface to cause visible surf; may be a hazard to ships (from the Arabic for "unborn").

- Microatoll: Community of species of corals; vertical growth limited by average tidal height; growth morphologies offer a low-resolution record of patterns of sea level change; fossilized remains can be dated using radioactive carbon dating and have been used to reconstruct Holocene sea levels.

- Cays: Small, low-elevation, sandy islands formed on the surface of coral reefs from eroded material that piles up, forming an area above sea level; can be stabilized by plants to become habitable; occur in tropical environments throughout the Pacific, Atlantic and Indian Oceans (including the Caribbean and on the Great Barrier Reef and Belize Barrier Reef), where they provide habitable and agricultural land.

- Seamount or guyot: Formed when a coral reef on a volcanic island subsides; tops of seamounts are rounded and guyots are flat; flat tops of guyots, or tablemounts, are due to erosion by waves, winds, and atmospheric processes.

Zones

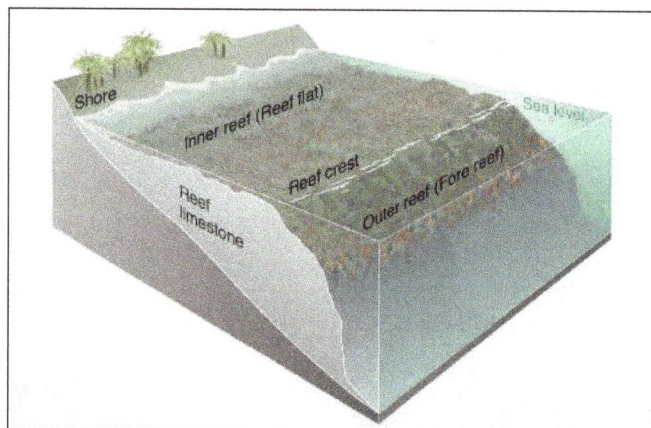

The three major zones of a coral reef: the fore reef, reef crest, and the back reef.

Coral reef ecosystems contain distinct zones that host different kinds of habitats. Usually, three major zones are recognized: the fore reef, reef crest, and the back reef (frequently referred to as the reef lagoon).

The three zones are physically and ecologically interconnected. Reef life and oceanic processes create opportunities for exchange of seawater, sediments, nutrients and marine life.

Most coral reefs exist in waters less than 50 m deep. Some inhabit tropical continental shelves where cool, nutrient-rich upwelling does not occur, such as the Great Barrier Reef. Others are found in the deep ocean surrounding islands or as atolls, such as in the Maldives. The reefs surrounding islands form when islands subside into the ocean, and atolls form when an island subsides below the surface of the sea.

Alternatively, Moyle and Cech distinguish six zones, though most reefs possess only some of the zones.

Water in the reef surface zone is often agitated. This diagram represents a reef on a continental shelf. The water waves at the left travel over the off-reef floor until they encounter the reef slope or fore reef. Then the waves pass over the shallow reef crest. When a wave enters shallow water it shoals, that is, it slows down and the wave height increases.

The reef surface is the shallowest part of the reef. It is subject to surge and tides. When waves pass over shallow areas, they shoal, as shown in the adjacent diagram. This means the water is often agitated. These are the precise condition under which corals flourish. The light is sufficient for photosynthesis by the symbiotic zooxanthellae, and agitated water brings plankton to feed the coral.

The off-reef floor is the shallow sea floor surrounding a reef. This zone occurs next to reefs on continental shelves. Reefs around tropical islands and atolls drop abruptly to great depths, and do not have such a floor. Usually sandy, the floor often supports seagrass meadows which are important foraging areas for reef fish.

The reef drop-off is, for its first 50 m, habitat for reef fish who find shelter on the cliff face and plankton in the water nearby. The drop-off zone applies mainly to the reefs surrounding oceanic islands and atolls.

The reef face is the zone above the reef floor or the reef drop-off. This zone is often the reef's most diverse area. Coral and calcareous algae provide complex habitats and areas that offer protection, such as cracks and crevices. Invertebrates and epiphytic algae provide much of the food for other organisms. A common feature on this forereef zone is spur and groove formations that serve to transport sediment downslope.

The reef flat is the sandy-bottomed flat, which can be behind the main reef, containing chunks of coral. This zone may border a lagoon and serve as a protective area, or it may lie between the reef and the shore, and in this case is a flat, rocky area. Fish tend to prefer it when it is present.

The reef lagoon is an entirely enclosed region, which creates an area less affected by wave action and often contains small reef patches.

However, the "topography of coral reefs is constantly changing. Each reef is made up of irregular patches of algae, sessile invertebrates, and bare rock and sand. The size, shape and relative abundance of these patches changes from year to year in response to the various factors that favor one type of patch over another. Growing coral, for example, produces constant change in the fine structure of reefs. On a larger scale, tropical storms may knock out large sections of reef and cause boulders on sandy areas to move."

Locations

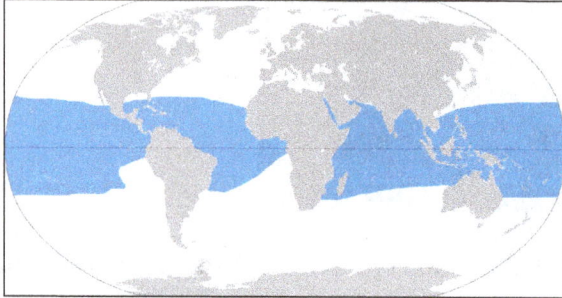

Boundary for 20 °C isotherms. Most corals live within this boundary. Note the cooler waters caused by upwelling on the southwest coast of Africa and off the coast of Peru.

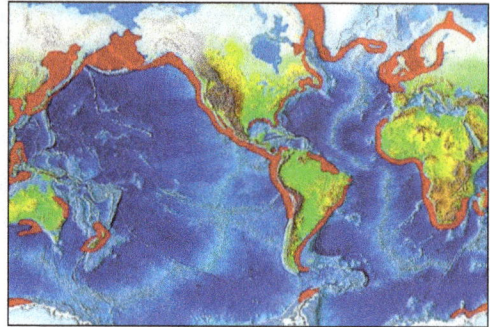

This map shows areas of upwelling in red. Coral reefs are not found in coastal areas where colder and nutrient-rich upwellings occur.

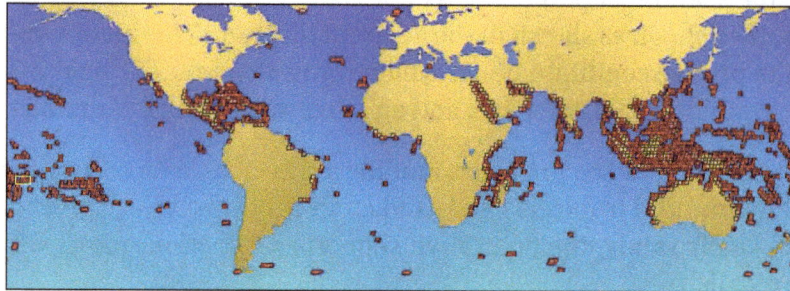

Locations of coral reefs.

Coral reefs are estimated to cover 284,300 km² (109,800 sq mi), just under 0.1% of the oceans' surface area. The Indo-Pacific region (including the Red Sea, Indian Ocean, Southeast Asia and the Pacific) account for 91.9% of this total. Southeast Asia accounts for 32.3% of that figure, while the Pacific including Australia accounts for 40.8%. Atlantic and Caribbean coral reefs account for 7.6%.

Although corals exist both in temperate and tropical waters, shallow-water reefs form only in a zone extending from approximately 30° N to 30° S of the equator. Tropical corals do not grow at depths of over 50 meters (160 ft). The optimum temperature for most coral reefs is 26–27 °C (79–81 °F), and few reefs exist in waters below 18 °C (64 °F). However, reefs in the Persian Gulf have adapted to temperatures of 13 °C (55 °F) in winter and 38 °C (100 °F) in summer. 37 species of scleractinian corals inhabit such an environment around Larak Island.

Deep-water coral inhabits greater depths and colder temperatures at much higher latitudes, as far north as Norway. Although deep water corals can form reefs, little is known about them.

Coral reefs are rare along the west coasts of the Americas and Africa, due primarily to upwelling and strong cold coastal currents that reduce water temperatures in these areas (the Peru, Benguela and Canary Currents respectively). Corals are seldom found along the coastline of South Asia from the eastern tip of India (Chennai) to the Bangladesh and Myanmar borders—as well as along the coasts of northeastern South America and Bangladesh, due to the freshwater release from the Amazon and Ganges Rivers respectively.

- The Great Barrier Reef—largest, comprising over 2,900 individual reefs and 900 islands stretching for over 2,600 kilometers (1,600 mi) off Queensland, Australia.

- The Mesoamerican Barrier Reef System—second largest, stretching 1,000 kilometers (620 mi) from Isla Contoy at the tip of the Yucatán Peninsula down to the Bay Islands of Honduras.

- The New Caledonia Barrier Reef—second longest double barrier reef, covering 1,500 kilometers (930 mi).

- The Andros, Bahamas Barrier Reef—third largest, following the east coast of Andros Island, Bahamas, between Andros and Nassau.

- The Red Sea—includes 6,000-year-old fringing reefs located along a 2,000 km (1,240 mi) coastline.

- The Florida Reef Tract—largest continental US reef and the third largest coral barrier reef, extends from Soldier Key, located in Biscayne Bay, to the Dry Tortugas in the Gulf of Mexico.

- Pulley Ridge—deepest photosynthetic coral reef, Florida.

- Numerous reefs around the Maldives.

- The Philippines coral reef area, the second largest in Southeast Asia, is estimated at 26,000 square kilometers. 915 reef fish species and more than 400 scleractinian coral species, 12 of which are endemic are found there.

- The Raja Ampat Islands in Indonesia's West Papua province offer the highest known marine diversity.

- Bermuda is known for its northernmost coral reef system, located at 32°24′N 64°48′W 32.4°N 64.8°W. The presence of coral reefs at this high latitude is due to the proximity of the Gulf Stream. Bermuda coral species represent a subset of those found in the greater Caribbean.

- The world's northernmost individual coral reef is located within a bay of Japan's Tsushima Island in the Korea Strait.

- The world's southernmost coral reef is at Lord Howe Island, in the Pacific Ocean off the east coast of Australia.

Coral

When alive, corals are colonies of small animals embedded in calcium carbonate shells. Coral heads consist of accumulations of individual animals called polyps, arranged in diverse shapes. Polyps are usually tiny, but they can range in size from a pinhead to 12 inches (30 cm) across.

Reef-building or hermatypic corals live only in the photic zone (above 50 m), the depth to which sufficient sunlight penetrates the water.

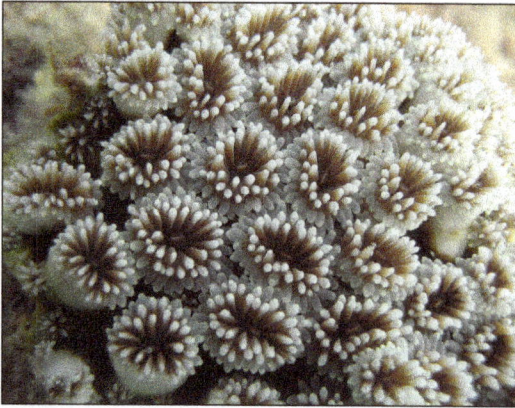

Close up of polyps arrayed on a coral, waving their tentacles. There can be thousands of polyps on a single coral branch.

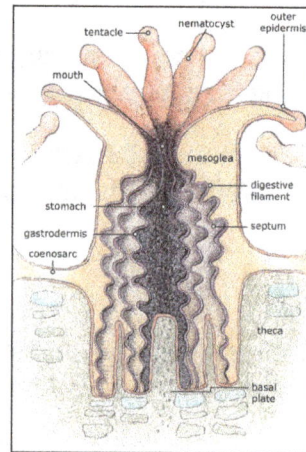

Diagram of a coral polyp anatomy.

Zooxanthellae

Coral polyps do not photosynthesize, but have a symbiotic relationship with microscopic algae (dinoflagellates) of the genus *Symbiodinium*, commonly referred to as zooxanthellae. These organisms live within the polyps' tissues and provide organic nutrients that nourish the polyp in the form of glucose, glycerol and amino acids. Because of this relationship, coral reefs grow much faster in clear water, which admits more sunlight. Without their symbionts, coral growth would be too slow to form significant reef structures. Corals get up to 90% of their nutrients from their symbionts. In return, as an example of mutualism, the corals shelter the zooxanthellae, averaging one million for every cubic centimeter of coral, and provide a constant supply of the carbon dioxide they need for photosynthesis.

The varying pigments in different species of zooxanthellae give them an overall brown or golden-brown appearance, and give brown corals their colors. Other pigments such as reds, blues, greens, etc. come from colored proteins made by the coral animals. Coral that loses a large fraction of its zooxanthellae becomes white (or sometimes pastel shades in corals that are pigmented with their own proteins) and is said to be bleached, a condition which, unless corrected, can kill the coral.

There are eight clades of *Symbiodinium* phylotypes. Most research has been conducted on clades A–D. Each clade contributes their own benefits as well as less compatible attributes to the survival of their coral hosts. Each photosynthetic organism has a specific level of sensitivity to photodamage to compounds needed for survival, such as proteins. Rates of regeneration and replication determine the organism's ability to survive. Phylotype A is found more in the shallow waters. It is able to produce mycosporine-like amino acids that are UV resistant, using a derivative of glycerin to absorb the UV radiation and allowing them to better adapt to warmer water temperatures. In the event of UV or thermal damage, if and when repair occurs, it will increase the likelihood of survival of the host and symbiont. This leads to the idea that, evolutionarily, clade A is more UV resistant and thermally resistant than the other clades.

Clades B and C are found more frequently in deeper water, which may explain their higher vulnerability to increased temperatures. Terrestrial plants that receive less sunlight because they are found in the undergrowth are analogous to clades B, C, and D. Since clades B through D are found at deeper depths, they require an elevated light absorption rate to be able to synthesize as much energy. With elevated absorption rates at UV wavelengths, these phylotypes are more prone to coral bleaching versus the shallow clade A.

Clade D has been observed to be high temperature-tolerant, and has a higher rate of survival than clades B and C during modern bleaching events.

Skeleton

Reefs grow as polyps and other organisms deposit calcium carbonate, the basis of coral, as a skeletal structure beneath and around themselves, pushing the coral head's top upwards and outwards. Waves, grazing fish (such as parrotfish), sea urchins, sponges and other forces and organisms act as bioeroders, breaking down coral skeletons into fragments that settle into spaces in the reef structure or form sandy bottoms in associated reef lagoons.

Table coral.

Typical shapes for coral species are named by their resemblance to terrestrial objects such as wrinkled brains, cabbages, table tops, antlers, wire strands and pillars. These shapes can depend on the life history of the coral, like light exposure and wave action, and events such as breakages.

Reproduction

Corals reproduce both sexually and asexually. An individual polyp uses both reproductive modes within its lifetime. Corals reproduce sexually by either internal or external fertilization. The reproductive cells are found on the mesenteries, membranes that radiate inward from the layer of tissue that lines the stomach cavity. Some mature adult corals are hermaphroditic; others are exclusively male or female. A few species change sex as they grow.

Internally fertilized eggs develop in the polyp for a period ranging from days to weeks. Subsequent development produces a tiny larva, known as a planula. Externally fertilized eggs develop during synchronized spawning. Polyps across a reef simultaneously release eggs and sperm into the water en masse. Spawn disperse over a large area. The timing of spawning depends on time of year, water temperature, and tidal and lunar cycles. Spawning is most successful given little variation between high and low tide. The less water movement, the better the chance for fertilization. Ideal timing occurs in the spring. Release of eggs or planula usually occurs at night, and is sometimes in phase with the lunar cycle (three to six days after a full moon). The period from release to settlement lasts only a few days, but some planulae can survive afloat for several weeks. They are vulnerable to predation and environmental conditions. The lucky few planulae that successfully attach to substrate then compete for food and space.

Other Reef Builders

Corals are the most prodigious reef-builders. However many other organisms living in the reef community contribute skeletal calcium carbonate in the same manner as corals. These include coralline algae and some sponges. Reefs are always built by the combined efforts of these different phyla, with different organisms leading reef-building in different geological periods.

Coralline Algae

Corraline algae Lithothamnion sp.

Coralline algae are important contributors to reef structure. Although their mineral deposition-rates are much slower than corals, they are more tolerant of rough wave-action, and so help to create a protective crust over those parts of the reef subjected to the greatest forces by waves, such as the reef front facing the open ocean. They also strengthen the reef structure by depositing limestone in sheets over the reef surface.

Sponges

Deep-water cloud sponge.

"Sclerosponge" is the descriptive name for all *Porifera* that build reefs. In the early Cambrian period, *Archaeocyatha* sponges were the world's first reef-building organisms, and sponges were the only reef-builders until the Ordovician. Sclerosponges still assist corals building modern reefs, but like coralline algae are much slower-growing than corals and their contribution is (usually) minor.

In the northern Pacific Ocean cloud sponges still create deep-water mineral-structures without corals, although the structures are not recognizable from the surface like tropical reefs. They are the only extant organisms known to build reef-like structures in cold water.

Gallery of Reef-Building Corals and their Reef-Building Assistants

Darwin's Paradox

Coralline algae corallina officinalis.

Staghorn coral.

"Coral seems to proliferate when ocean waters are warm, poor, clear and agitated, a fact which Darwin had already noted when he passed through Tahiti in 1842. This constitutes a fundamental paradox, shown quantitatively by the apparent impossibility of balancing input and output of the nutritive elements which control the coral polyp metabolism.

Recent oceanographic research has brought to light the reality of this paradox by confirming that the oligotrophy of the ocean euphotic zone persists right up to the swell-battered reef crest. When you approach the reef edges and atolls from the quasidesert of the open sea, the near absence of living matter suddenly becomes a plethora of life, without transition. So why is there something rather than nothing, and more precisely, where do the necessary nutrients for the functioning of this extraordinary coral reef machine come from?"

In *The Structure and Distribution of Coral Reefs*, published in 1842, Darwin described how coral reefs were found in some tropical areas but not others, with no obvious cause. The largest and strongest corals grew in parts of the reef exposed to the most violent surf and corals were weakened or absent where loose sediment accumulated.

Tropical waters contain few nutrients yet a coral reef can flourish like an "oasis in the desert". This has given rise to the ecosystem conundrum, sometimes called "Darwin's paradox": "How can such high production flourish in such nutrient poor conditions?"

Coral reefs support over one-quarter of all marine species. This diversity results in complex food webs, with large predator fish eating smaller forage fish that eat yet smaller zooplankton and so on. However, all food webs eventually depend on plants, which are the primary producers. Coral reefs typically produce 5–10 grams of carbon per square meter per day ($gC \cdot m^{-2} \cdot day^{-1}$) biomass.

One reason for the unusual clarity of tropical waters is their nutrient deficiency and drifting plankton. Further, the sun shines year-round in the tropics, warming the surface layer, making it less dense than subsurface layers. The warmer water is separated from deeper, cooler water by a stable thermocline, where the temperature makes a rapid change. This keeps the warm surface waters floating above the cooler deeper waters. In most parts of the ocean, there is little exchange between these layers. Organisms that die in aquatic environments generally sink to the bottom, where they decompose, which releases nutrients in the form of nitrogen (N), phosphorus (P) and potassium (K). These nutrients are necessary for plant growth, but in the tropics, they do not directly return to the surface.

Plants form the base of the food chain and need sunlight and nutrients to grow. In the ocean, these plants are mainly microscopic phytoplankton which drift in the water column. They need sunlight for photosynthesis, which powers carbon fixation, so they are found only relatively near the surface, but they also need nutrients. Phytoplankton rapidly use nutrients in the surface waters, and in the tropics, these nutrients are not usually replaced because of the thermocline.

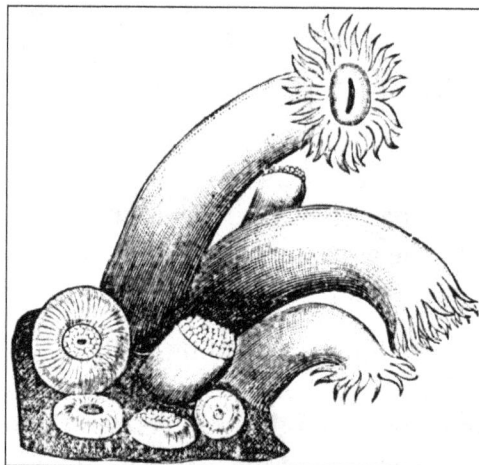

Coral polyps.

Explanations

Around coral reefs, lagoons fill in with material eroded from the reef and the island. They become havens for marine life, providing protection from waves and storms.

Most importantly, reefs recycle nutrients, which happens much less in the open ocean. In coral reefs and lagoons, producers include phytoplankton, as well as seaweed and coralline algae, especially small types called turf algae, which pass nutrients to corals. The phytoplankton form the base of the food chain and are eaten by fish and crustaceans. Recycling reduces the nutrient inputs needed overall to support the community.

The color of corals depends on the combination of brown shades provided by their zooxanthellae
and pigmented proteins (reds, blues, greens, etc.) produced by the corals themselves.

Corals also absorb nutrients, including inorganic nitrogen and phosphorus, directly from water.
Many corals extend their tentacles at night to catch zooplankton that pass near. Zooplankton provide the polyp with nitrogen, and the polyp shares some of the nitrogen with the zooxanthellae,
which also require this element.

Sponges live in crevices in the reefs. They are efficient filter feeders, and in the Red Sea they consume about 60% of the phytoplankton that drifts by. Sponges eventually excrete nutrients in a
form that corals can use.

Most coral polyps are nocturnal feeders. Here, in the dark, polyps have extended their
tentacles to feed on zooplankton.

The roughness of coral surfaces is key to coral survival in agitated waters. Normally, a boundary
layer of still water surrounds a submerged object, which acts as a barrier. Waves breaking on the
extremely rough edges of corals disrupt the boundary layer, allowing the corals access to passing
nutrients. Turbulent water thereby promotes reef growth. Without the access to nutrients brought
by rough coral surfaces, even the most effective recycling would not suffice.

Deep nutrient-rich water entering coral reefs through isolated events may have significant effects
on temperature and nutrient systems. This water movement disrupts the relatively stable thermocline that usually exists between warm shallow water and deeper colder water. Temperature
regimes on coral reefs in the Bahamas and Florida are highly variable with temporal scales of minutes to seasons and spatial scales across depths.

Water can pass through coral reefs in various ways, including current rings, surface waves, internal waves and tidal changes. Movement is generally created by tides and wind. As tides interact with varying bathymetry and wind mixes with surface water, internal waves are created. An internal wave is a gravity wave that moves along density stratification within the ocean. When a water parcel encounters a different density it oscillates and creates internal waves. While internal waves generally have a lower frequency than surface waves, they often form as a single wave that breaks into multiple waves as it hits a slope and moves upward. This vertical breakup of internal waves causes significant diapycnal mixing and turbulence. Internal waves can act as nutrient pumps, bringing plankton and cool nutrient-rich water to the surface.

The irregular structure characteristic of coral reef bathymetry may enhance mixing and produce pockets of cooler water and variable nutrient content. Arrival of cool, nutrient-rich water from depths due to internal waves and tidal bores has been linked to growth rates of suspension feeders and benthic algae as well as plankton and larval organisms. The seaweed *Codium isthmocladum* reacts to deep water nutrient sources because their tissues have different concentrations of nutrients dependent upon depth. Aggregations of eggs, larval organisms and plankton on reefs respond to deep water intrusions. Similarly, as internal waves and bores move vertically, surface-dwelling larval organisms are carried toward the shore. This has significant biological importance to cascading effects of food chains in coral reef ecosystems and may provide yet another key to unlocking the paradox.

Cyanobacteria provide soluble nitrates via nitrogen fixation.

Coral reefs often depend on surrounding habitats, such as seagrass meadows and mangrove forests, for nutrients. Seagrass and mangroves supply dead plants and animals that are rich in nitrogen and serve to feed fish and animals from the reef by supplying wood and vegetation. Reefs, in turn, protect mangroves and seagrass from waves and produce sediment in which the mangroves and seagrass can root.

Biodiversity

Tube sponges attracting cardinal fishes, glassfishes and wrasses.

Over 4,000 species of fish inhabit coral reefs.

Organisms can cover every square inch of a coral reef.

Coral reefs form some of the world's most productive ecosystems, providing complex and varied marine habitats that support a wide range of other organisms. Fringing reefs just below low tide level have a mutually beneficial relationship with mangrove forests at high tide level and sea grass meadows in between: the reefs protect the mangroves and seagrass from strong currents and waves that would damage them or erode the sediments in which they are rooted, while the mangroves and sea grass protect the coral from large influxes of silt, fresh water and pollutants. This level of variety in the environment benefits many coral reef animals, which, for example, may feed in the sea grass and use the reefs for protection or breeding.

Reefs are home to a variety of animals, including fish, seabirds, sponges, cnidarians (which includes some types of corals and jellyfish), worms, crustaceans (including shrimp, cleaner shrimp, spiny lobsters and crabs), mollusks (including cephalopods), echinoderms (including starfish, sea urchins and sea cucumbers), sea squirts, sea turtles and sea snakes. Aside from humans, mammals are rare on coral reefs, with visiting cetaceans such as dolphins the main exception. A few species feed directly on corals, while others graze on algae on the reef. Reef biomass is positively related to species diversity.

The same hideouts in a reef may be regularly inhabited by different species at different times of day. Nighttime predators such as cardinalfish and squirrelfish hide during the day, while damselfish, surgeonfish, triggerfish, wrasses and parrotfish hide from eels and sharks.

The great number and diversity of hiding places in coral reefs, i.e. refuges, are the most important factor causing the great diversity and high biomass of the organisms in coral reefs.

Algae

Reefs are chronically at risk of algal encroachment. Overfishing and excess nutrient supply from onshore can enable algae to outcompete and kill the coral. Increased nutrient levels can be a result of sewage or chemical fertilizer runoff. Runoff can carry nitrogen and phosphorus which promote excess algae growth. Algae can sometimes out-compete the coral for space. The algae can then smother the coral by decreasing the oxygen supply available to the reef. Decreased oxygen levels can slow down calcification rates, weakening the coral and leaving it more susceptible to disease and degradation. Algae inhabit a large percentage of surveyed coral locations. The algal population consists of turf algae, coralline algae and macro algae. Some sea urchins (such as *Diadema antillarum*) eat these algae and could thus decrease the risk of algal encroachment.

Sponges

Sponges are essential for the functioning of the coral reef that system. Algae and corals in coral reefs produce organic material. This is filtered through sponges which convert this organic material into small particles which in turn are absorbed by algae and corals.

Fish

Over 4,000 species of fish inhabit coral reefs. The reasons for this diversity remain unclear. Hypotheses include the "lottery", in which the first (lucky winner) recruit to a territory is typically able to defend it against latecomers, "competition", in which adults compete for territory, and less-competitive species must be able to survive in poorer habitat, and "predation", in which population size is a function of postsettlement piscivore mortality. Healthy reefs can produce up to 35 tons of fish per square kilometer each year, but damaged reefs produce much less.

Invertebrates

Sea urchins, *Dotidae* and sea slugs eat seaweed. Some species of sea urchins, such as *Diadema antillarum*, can play a pivotal part in preventing algae from overrunning reefs. Researchers are investigating the use of native collector urchins, *Tripneustes gratilla*, for their potential as biocontrol agents to mitigate the spread of invasive algae species on coral reefs. *Nudibranchia* and sea anemones eat sponges.

A number of invertebrates, collectively called "cryptofauna," inhabit the coral skeletal substrate itself, either boring into the skeletons (through the process of bioerosion) or living in pre-existing voids and crevices. Animals boring into the rock include sponges, bivalve mollusks, and sipunculans. Those settling on the reef include many other species, particularly crustaceans and polychaete worms.

Seabirds

Coral reef systems provide important habitats for seabird species, some endangered. For example, Midway Atoll in Hawaii supports nearly three million seabirds, including two-thirds (1.5 million) of the global population of Laysan albatross, and one-third of the global population of black-footed albatross. Each seabird species has specific sites on the atoll where they nest. Altogether, 17 species of seabirds live on Midway. The short-tailed albatross is the rarest, with fewer than 2,200 surviving after excessive feather hunting in the late 19th century.

Other

Sea snakes feed exclusively on fish and their eggs. Marine birds, such as herons, gannets, pelicans and boobies, feed on reef fish. Some land-based reptiles intermittently associate with reefs, such as monitor lizards, the marine crocodile and semiaquatic snakes, such as *Laticauda colubrina*. Sea turtles, particularly hawksbill sea turtles, feed on sponges.

Schooling reef fish.

Caribbean reef squid.

Ecosystem Services

Coral reefs deliver ecosystem services to tourism, fisheries and coastline protection. The global economic value of coral reefs has been estimated to be between US $29.8 billion and $375 billion per year.

The economic cost over a 25-year period of destroying one kilometer of coral reef has been estimated to be somewhere between $137,000 and $1,200,000.

To improve the management of coastal coral reefs, the World Resources Institute (WRI) developed and published tools for calculating the value of coral reef-related tourism, shoreline protection and fisheries, partnering with five Caribbean countries. As of April 2011, published working papers covered St. Lucia, Tobago, Belize, and the Dominican Republic. The WRI was "making sure that the study results support improved coastal policies and management planning". The Belize study estimated the value of reef and mangrove services at $395–559 million annually.

Bermuda's coral reefs provide economic benefits to the Island worth on average $722 million per year, based on six key ecosystem services.

Shoreline Protection

Coral reefs protect shorelines by absorbing wave energy, and many small islands would not exist without reefs. Coral reefs can reduce wave energy by 97%, helping to prevent loss of life and property damage. Coastlines protected by coral reefs are also more stable in terms of erosion than those without. Reefs can attenuate waves as well as or better than artificial structures designed for coastal defence such as breakwaters. An estimated 197 million people who live both below 10 m elevation and within 50 km of a reef consequently may receive risk reduction benefits from reefs. Restoring reefs is significantly cheaper than building artificial breakwaters in tropical environments. Expected damages from flooding would double, and costs from frequent storms would triple without the topmost meter of reefs. For 100-year storm events, flood damages would increase by 91% to $US 272 billion without the top meter.

Fisheries

About six million tons of fish are taken each year from coral reefs. Well-managed reefs have an average annual yield of 15 tons of seafood per square kilometer. Southeast Asia's coral reef fisheries alone yield about $2.4 billion annually from seafood.

Threats

Coral reefs are dying around the world. In particular, runoff, pollution, overfishing, blast fishing, disease, invasive species, overuse by humans and coral mining and the digging of canals and access into islands and bays are localized threats to coral ecosystems. Broader threats are sea temperature rise, sea level rise and ocean acidification, all associated with greenhouse gas emissions. Other threats include the ocean's role as a carbon dioxide sink, atmospheric changes, ultraviolet light, ocean acidification, viruses, impacts of dust storms carrying agents to far-flung reefs, and algal blooms.

Air pollution can stunt the growth of coral reefs; including coal-burning and volcanic eruptions. Pollutants, such as Tributyltin, a biocide released into water from anti-fouling paint can be toxic to corals.

In 2011, researchers suggested that "extant marine invertebrates face the same synergistic effects of multiple stressors" that occurred during the end-Permian extinction, and that genera "with poorly buffered respiratory physiology and calcareous shells", such as corals, were particularly vulnerable.

Rock coral on seamounts are threatened by bottom trawling. Reportedly up to 50% of the catch is rock coral, and the practice smashes coral structures to rubble. These ecosystems take years to regrow, destroying coral communities faster than they can rebuild.

Another cause for the death of coral reefs is bioerosion. Various fishes graze corals and change the morphology of coral reefs making them more susceptible to other threats. Only the algae growing on dead corals is eaten and the live ones are not. However, this act still destroys the top layer of coral substrate and makes it harder for the reefs to sustain.

In El Niño-year 2010, global coral bleaching reached its worst level since El Niño year 1998, when 16% of the world's reefs died as a result of increased water temperature. In Indonesia's Aceh province, surveys showed some 80% of bleached corals died. Bleaching leaves corals vulnerable to disease, stunts their growth, and affects their reproduction, while severe bleaching kills them. In July, Malaysia closed several dive sites where virtually all the corals were damaged by bleaching.

Coral reefs with one type of zooxanthellae are more prone to bleaching than are reefs with another, more hardy, species.

Ecotourism in the Great Barrier Reef contributes to coral disease. Chemicals in sunscreens may contribute to the impact of viruses on zooxanthellae and impact reproduction.

In a large-scale systematic study of Jarvis Island coral community, scientists have observed ten coral bleaching events from 1960 to 2016.

Protection

A diversity of corals.

Marine protected areas (MPAs) are designated areas that provide various kinds of protection to ocean and/or estuarine areas. They are intended to promote responsible fishery management and habitat protection. MPAs can encompass both social and biological objectives, including reef restoration, aesthetics, biodiversity and economic benefits.

However, research in Indonesia, Philippines and Papua New Guinea found no significant difference between an MPA site and an unprotected site. Further, they can generate conflicts driven by lack of community participation, clashing views of the government and fisheries, effectiveness of the area and funding. In some situations, as in the Phoenix Islands Protected Area, MPAs provide revenue, potentially equal to the income they would have generated without controls.

According to the Caribbean Coral Reefs - Status Report 19702-2012, states that; stop overfishing especially fishes key to coral reef like parrotfish, coastal zone management that reduce human pressure on reef, (for example restricting coastal settlement, development and tourism) and control pollution specially sewage, may reduce coral decline or even reverse it. The report shows that healthier reefs in the Caribbean are those with large populations of parrotfish in countries that protect these key fishes and sea urchins, banning fish trapping and spearfishing, creating "resilient reefs".

To help combat ocean acidification, some laws are in place to reduce greenhouse gases such as carbon dioxide. The United States Clean Water Act puts pressure on state governments to monitor and limit runoff.

Many land use laws aim to reduce CO_2 emissions by limiting deforestation. Deforestation can release significant amounts of CO_2 absent sequestration via active follow-up forestry programs. Deforestation can also cause erosion, which flows into the ocean, contributing to ocean acidification. Incentives are used to reduce miles traveled by vehicles, which reduces carbon emissions into the atmosphere, thereby reducing the amount of dissolved CO_2 in the ocean. State and federal governments also regulate land activities that affect coastal erosion. High-end satellite technology can monitor reef conditions.

Designating a reef as a biosphere reserve, marine park, national monument or world heritage site can offer protections. For example, Belize's barrier reef, Sian Ka'an, the Galapagos islands, Great Barrier Reef, Henderson Island, Palau and Papahānaumokuākea Marine National Monument are world heritage sites.

In Australia, the Great Barrier Reef is protected by the Great Barrier Reef Marine Park Authority, and is the subject of much legislation, including a biodiversity action plan. Australia compiled a Coral Reef Resilience Action Plan. This plan consists of adaptive management strategies, including reducing carbon footprint. A public awareness plan providezs education on the "rainforests of the sea" and how people can reduce carbon emissions.

Inhabitants of Ahus Island, Manus Province, Papua New Guinea, have followed a generations-old practice of restricting fishing in six areas of their reef lagoon. Their cultural traditions allow line fishing, but no net or spear fishing. Both biomass and individual fish sizes are significantly larger than in places where fishing is unrestricted.

Restoration

Coral Farming

Coral aquaculture, also known as coral farming or coral gardening, is showing promise as a potentially effective tool for restoring coral reefs.

The "gardening" process bypasses the early growth stages of corals when they are most at risk of dying. Coral seeds are grown in nurseries, then replanted on the reef. Coral is farmed by coral farmers whose interests range from reef conservation to increased income.

Coral fragments growing on nontoxic concrete.

Creating Substrate

Efforts to expand the size and number of coral reefs generally involve supplying substrate to allow more corals to find a home. Substrate materials include discarded vehicle tires, scuttled ships, subway cars and formed concrete, such as reef balls. Reefs grow unaided on marine structures such as oil rigs. In large restoration projects, propagated hermatypic coral on substrate can be secured with metal pins, superglue or milliput. Needle and thread can also attach A-hermatype coral to substrate.

Biorock is a substrate produced by a patented process that runs low voltage electrical currents through seawater to cause dissolved minerals to precipitate onto steel structures. The resultant white carbonate (aragonite) is the same mineral that makes up natural coral reefs. Corals rapidly colonize and grow at accelerated rates on these coated structures. The electrical currents also accelerate formation and growth of both chemical limestone rock and the skeletons of corals and other shell-bearing organisms, such as oysters. The vicinity of the anode and cathode provides a high-pH environment which inhibits the growth of competitive filamentous and fleshy algae. The increased growth rates fully depend on the accretion activity.

Under the influence of the electric field, corals display an increased growth rate, size and density.

Relocation

One case study with coral reef restoration was conducted on the island of Oahu in Hawaii. The University of Hawaii operates a Coral Reef Assessment and Monitoring Program to help relocate and restore coral reefs in Hawaii. A boat channel from the island of Oahu to the Hawaii Institute of Marine Biology on Coconut Island was overcrowded with coral reefs. Many areas of coral reef patches in the channel had been damaged from past dredging in the channel.

Dredging covers corals with sand. Coral larvae cannot settle on sand; they can only build on existing reefs or compatible hard surfaces, such as rock or concrete. Because of this, the University decided to relocate some of the coral. They transplanted them with the help of United States Army divers, to a site relatively close to the channel. They observed little if any damage to any of the colonies during transport and no mortality of coral reefs was observed on the transplant site. While attaching the coral to the transplant site, they found that coral placed on hard rock grew well, including on the wires that attached the corals to the site.

No environmental effects were seen from the transplantation process, recreational activities were not decreased, and no scenic areas were affected.

Heat-tolerant Symbionts

Another possibility for coral restoration is gene therapy: inoculating coral with genetically modified bacteria, or naturally-occurring heat-tolerant varieties of coral symbiotes, may make it possible to grow corals that are more resistant to climate change and other threats.

Invasive Algae

Hawaiian coral reefs smothered by the spread of invasive algae were managed with a two-prong approach: divers manually removed invasive algae, with the support of super-sucker barges. Grazing pressure on invasive algae needed to be increased to prevent the regrowth of the algae.

Researchers found that native collector urchins were reasonable candidate grazers for algae bio-control, to extirpate the remaining invasive algae from the reef.

Pelagic Zone

The pelagic zone consists of the water column of the open ocean, and can be further divided into regions by depth. The pelagic zone can be thought of in terms of an imaginary cylinder or water column that goes from the surface of the sea almost to the bottom. Conditions differ deeper in the water column such that as pressure increases with depth, the temperature drops and less light penetrates. Depending on the depth, the water column, rather like the Earth's atmosphere, may be divided into different layers.

The pelagic zone occupies 1,330 million km³ (320 million mi³) with a mean depth of 3.68 km (2.29 mi) and maximum depth of 11 km (6.8 mi). Fish that live in the pelagic zone are called pelagic fish. Pelagic life decreases with increasing depth. It is affected by light intensity, pressure, temperature, salinity, the supply of dissolved oxygen and nutrients, and the submarine topography, which is called bathymetry. In deep water, the pelagic zone is sometimes called the open-ocean zone and can be contrasted with water that is near the coast or on the continental shelf. In other contexts, coastal water not near the bottom is still said to be in the pelagic zone.

The pelagic zone can be contrasted with the benthic and demersal zones at the bottom of the sea. The benthic zone is the ecological region at the very bottom of the sea. It includes the sediment surface and some subsurface layers. Marine organisms living in this zone, such as clams and crabs, are called benthos. The demersal zone is just above the benthic zone. It can be significantly affected by the seabed and the life that lives there. Fish that live in the demersal zone are called demersal fish, and can be divided into benthic fish, which are denser than water so they can rest on the bottom, and benthopelagic fish, which swim in the water column just above the bottom. Demersal fish are also known as bottom feeders and groundfish.

Depth and Layers

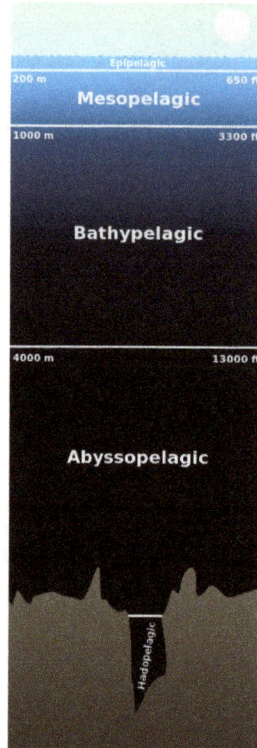

A scale diagram of the layers of the pelagic zone.

Depending on how deep the sea is, the pelagic zone can extend to five vertical regions in the ocean. From the top down, these are:

Epipelagic (Sunlight)

From the surface (MSL) down to around 200 m (660 ft).

This is the illuminated zone at the surface of the sea where enough light is available for photosynthesis. Nearly all primary production in the ocean occurs here. Consequently, plants and animals are largely concentrated in this zone. Examples of organisms living in this zone are plankton, floating seaweed, jellyfish, tuna, many sharks and dolphins.

Mesopelagic (Twilight)

From 200 m (660 ft) down to around 1,000 m (3,300 ft).

The most abundant organisms thriving into the mesopelagic zone are heterotrophic bacteria. Examples of animals that live here are swordfish, squid, *Anarhichadidae* or "wolffish" and some species of cuttlefish. Many organisms that live in this zone are bioluminescent. Some creatures living in the mesopelagic zone rise to the epipelagic zone at night to feed.

Bathypelagic (Midnight)

From 1,000 m (3,300 ft) down to around 4,000 m (13,000 ft).

At this depth, the ocean is pitch black, apart from occasional bioluminescent organisms, such as anglerfish. No living plant exists here. Most animals living here survive by consuming the detritus falling from the zones above, which is known as "marine snow", or, like the marine hatchetfish, by preying on other inhabitants of this zone. Other examples of this zone's inhabitants are giant squid, smaller squids and the grimpoteuthis or "dumbo octopus". The giant squid is hunted here by deep-diving sperm whales.

Abyssopelagic (Lower Midnight)

From around 4,000 m (13,000 ft) down to above the ocean floor.

Very few creatures live in the cold temperatures, high pressures and complete darkness of this depth. Among the species found in this zone are several species of squid; echinoderms including the basket star, swimming cucumber, and the sea pig; and marine arthropods including the sea spider. Many of the species living at these depths are transparent and eyeless because of the total lack of light in this zone.

Hadopelagic

This is the deepest part of the ocean at more than 6,000 m (20,000 ft) or 6,500 m (21,300 ft), depending on authority. Such depths are generally located in trenches.

Pelagic Ecosystem

The pelagic ecosystem is based on phytoplankton. Phytoplankton manufacture their own food using a process of photosynthesis. Because they need sunlight, they inhabit the upper, sunlit epipelagic zone, which includes the coastal or neritic zone. Biodiversity diminishes markedly in the deeper zones below the epipelagic zone as dissolved oxygen diminishes, water pressure increases, temperatures become colder, food sources become scarce, and light diminishes and finally disappears.

The pelagic sooty tern spends months at a time flying at sea, returning to land only for breeding.

Pelagic Birds

Pelagic birds, also called oceanic birds, live on the open sea, rather than around waters adjacent to land or around inland waters. Pelagic birds feed on planktonic crustaceans, squid and forage fish. Examples are the Atlantic puffin, macaroni penguins, sooty terns, shearwaters, and Procellariiformes such as the albatross, Procellariidae and petrels.

The term seabird includes birds which live around the sea adjacent to land, as well as pelagic birds.

Pelagic Fish

Pelagic fish live in the water column of coastal, ocean, and lake waters, but not on or near the bottom of the sea or the lake. They can be contrasted with demersal fish, which live on or near the bottom, and coral reef fish.

These fish are often migratory forage fish, which feed on plankton, and the larger fish that follow and feed on the forage fish. Examples of migratory forage fish are herring, anchovies, capelin, and menhaden. Examples of larger pelagic fish which prey on the forage fish are billfish, tuna, and oceanic sharks.

Pelagic Invertebrates

Some examples of pelagic invertebrates include krill, copepods, jellyfish, decapod larvae, hyperiid amphipods, rotifers and cladocerans.

Thorson's rule states that benthic marine invertebrates at low latitudes tend to produce large numbers of eggs developing to widely dispersing pelagic larvae, whereas at high latitudes such organisms tend to produce fewer and larger lecithotrophic (yolk-feeding) eggs and larger offspring.

Pelagic Reptiles

Pelamis platura, the pelagic sea snake, is the only one of the 65 species of marine snakes to spend its entire life in the pelagic zone. It bears live young at sea and is helpless on land. The species sometimes forms aggregations of thousands along slicks in surface waters. The pelagic sea snake is the world's most widely distributed snake species.

Many species of sea turtles spend the first years of their lives in the pelagic zone, moving closer to shore as they reach maturity.

Intertidal Zone

The intertidal zone, also known as the foreshore or seashore, is the area that is above water level at low tide and underwater at high tide (in other words, the area within the tidal range). This area can include several types of habitats with various species of life, such as starfish, sea urchins, and many species of coral. Sometimes it is referred to as the littoral zone, although that can be defined as a wider region.

Bancao Beach at low tide, showing the intertidal zone about 200 m from the beach.

The well-known area also includes steep rocky cliffs, sandy beaches, or wetlands (e.g., vast mud-flats). The area can be a narrow strip, as in Pacific islands that have only a narrow tidal range, or can include many meters of shoreline where shallow beach slopes interact with high tidal excursion. The peritidal zone is similar but somewhat wider, extending from above the highest tide level to below the lowest.

Organisms in the intertidal zone are adapted to an environment of harsh extremes.

The intertidal zone is also home to several species from different phyla (Porifera, Annelida, Coelenterata, Mollusca, Arthropoda, etc.). Water is available regularly with the tides, but varies from fresh with rain to highly saline and dry salt, with drying between tidal inundations. Wave splash can dislodge residents from the littoral zone. With the intertidal zone's high exposure to sunlight, the temperature can range from very hot with full sunshine to near freezing in colder climates. Some microclimates in the littoral zone are ameliorated by local features and larger plants such as mangroves. Adaptation in the littoral zone allows the use of nutrients supplied in high volume on a regular basis from the sea, which is actively moved to the zone by tides. Edges of habitats, in this case land and sea, are themselves often significant ecologies, and the littoral zone is a prime example.

A typical rocky shore can be divided into a spray zone or splash zone (also known as the supratidal zone), which is above the spring high-tide line and is covered by water only during storms, and an intertidal zone, which lies between the high and low tidal extremes. Along most shores, the intertidal zone can be clearly separated into the following subzones: high tide zone, middle tide zone, and low tide zone. The intertidal zone is one of a number of marine biomes or habitats, including estuary, neritic, surface, and deep zones.

Zonation

Marine biologists divide the intertidal region into three zones (low, middle, and high), based on the overall average exposure of the zone. The low intertidal zone, which borders on the shallow subtidal zone, is only exposed to air at the lowest of low tides and is primarily marine in character. The mid intertidal zone is regularly exposed and submerged by average tides. The high intertidal zone is only covered by the highest of the high tides, and spends much of its time as terrestrial habitat. The high intertidal zone borders on the splash zone (the region above the highest still-tide level, but which receives wave splash). On shores exposed to heavy wave action, the intertidal zone will be influenced by waves, as the spray from breaking waves will extend the intertidal zone.

Tide pools at Pillar Point showing zonation on the edge of the rock ledge.

A rock, seen at low tide, exhibiting typical intertidal zonation, Kalaloch, Washington, western United States.

Depending on the substratum and topography of the shore, additional features may be noticed. On rocky shores, tide pools form in depressions that fill with water as the tide rises. Under certain conditions, such as those at Morecambe Bay, quicksand may form.

Low Tide Zone (Lower Littoral)

This subregion is mostly submerged - it is only exposed at the point of low tide and for a longer period of time during extremely low tides. This area is teeming with life; the most notable difference with this subregion to the other three is that there is much more marine vegetation, especially seaweeds. There is also a great biodiversity. Organisms in this zone generally are not well adapted to periods of dryness and temperature extremes. Some of the organisms in this area are abalone, sea anemones, brown seaweed, chitons, crabs, green algae, hydroids, isopods, limpets, mussels, nudibranchs, sculpin, sea cucumber, sea lettuce, sea palms, starfish, sea urchins, shrimp, snails, sponges, surf grass, tube worms, and whelks. Creatures in this area can grow to larger sizes because there is more available energy in the localized ecosystem. Also, marine vegetation can grow to much greater sizes than in the other three intertidal subregions due to the better water coverage. The water is shallow enough to allow plenty of light to reach the vegetation to allow substantial photosynthetic activity, and the salinity is at almost normal levels. This area is also protected from large predators such as fish because of the wave action and the relatively shallow water.

Ecology

A California tide pool in the low tide zone.

The intertidal region is an important model system for the study of ecology, especially on wave-swept rocky shores. The region contains a high diversity of species, and the zonation created by the tides causes species ranges to be compressed into very narrow bands. This makes it relatively simple to study species across their entire cross-shore range, something that can be extremely difficult in, for instance, terrestrial habitats that can stretch thousands of kilometres. Communities on wave-swept shores also have high turnover due to disturbance, so it is possible to watch ecological succession over years rather than decades.

The burrowing invertebrates that make up large portions of sandy beach ecosystems are known to travel relatively great distances in cross-shore directions as beaches change on the order of days, semilunar cycles, seasons, or years. The distribution of some species has been found to correlate strongly with geomorphic datums such as the high tide strand and the water table outcrop.

Since the foreshore is alternately covered by the sea and exposed to the air, organisms living in this environment must have adaptions for both wet and dry conditions. Hazards include being smashed or carried away by rough waves, exposure to dangerously high temperatures, and desiccation. Typical inhabitants of the intertidal rocky shore include urchins, sea anemones, barnacles, chitons, crabs, isopods, mussels, starfish, and many marine gastropod molluscs such as limpets and whelks.

References

- Habitats: Beaches - Characteristics Archived 2011-05-26 at the Wayback Machine Office of Naval Research. Retrieved 17 April 2011

- Types-of-Marine-Organisms-HS-ES, lesson, types-of-marine-organisms, earth-science: ck12.org, Retrieved 22 July, 2019

- "7,000 m Class Remotely Operated Vehicle KAIKO 7000". Japan Agency for Marine-Earth Science and Technology (JAMSTEC). Retrieved 7 June 2008

- Rona, Peter A. (2003). "Resources of the Sea Floor". Science. 299 (5607): 673–674. Doi:10.1126/science.1080679. PMID 12560541. Retrieved 2007-02-04

- Marine-ecosystem, science: britannica.com, Retrieved 21 June 2019

- Birdlife International (2008). "Sterna fuscata". IUCN Red List of Threatened Species. Version 2008. International Union for Conservation of Nature. Retrieved 7 August 2009

- Cressey, Daniel (21 November 2007). "Giant sea scorpion discovered". Nature. Doi:10.1038/news.2007.272. Retrieved 10 June 2013

Threats to Marine Ecosystem

4

- **Primary Causes of Destruction of Marine Biological Diversity**
- **Habitat Alteration**
- **Toxic Pollution**
- **Invasive Species**
- **Marine Conservation**

There are numerous factors which threaten the health of marine ecosystems. Ocean dumping, land runoff, ocean acidification, pollution are some of the common threats to marine ecosystems. The topics elaborated in this will help in gaining a better perspective about these threats to marine ecosystems as well as the conservation practices.

Threats to Marine Ecosystems

It is true to say that despite taking numerous steps to mitigate the effects of marine pollution, there is still a long way to go to protect and conserve our waterbodies. As important as finding ways to clean our oceans and lakes, creating awareness among people towards the protection and conservation of the marine environment is also a crucial effort. And being aware of a problem means knowing the issue at the grass root level. Thus, to prevent marine pollution, one must be aware of the pollutants that pose threat to the ecosystem and the sources those originate.

Here, we have enumerated 11 main causes of marine pollution which have been a troubling marine environment for quite some time now.

Here, we have enumerated 11 main causes of marine pollution which have been a troubling marine environment for quite some time now.

Ocean Dumping

Dumping of waste materials from industries, ships and sewage plants into oceans has polluted the marine ecosystem to a great extent. As mentioned earlier, for years the oceans have been targeted as a dumping site of sewage, chemicals, industrial waste, garbage, and other debris from the land. According to reports, only mining companies across the world dump 220 million tonnes hazardous waste directly into our waterbodies every year. Similarly, it is important to note that around two-thirds of the marine lives in the world have been under threat from the chemicals we use every day, including household cleaners. Since we depend on the marine ecosystem extensively, the adverse effects of ocean dumping have not only been felt by marine life but by humans as it poses health risks.

Land Runoff

One of the major sources of ocean pollution is the waste comes from the nonpoint source, which happens as a result of runoff. Surface runoff from both agricultural land and areas carry soil and particles mixed with carbon, phosphorus, nitrogen and minerals, posing threat to the marine life in alarming scale. Crossing streams and rivers, the water filled with these toxic chemicals land in the ocean, resulting in harmful algal blooms. This kind of water pollution threatens the species of fish, turtles, and shrimp etc. and also humans through the food chain.

Dredging

In this world of ever expanding industrial activities, dredging is an important activity that enhances marine transportation and other related activities. However, dredging has been a major cause of disturbance in the marine ecosystem for many years. As dredging is to remove the deposits submerged underwater, the activity alters the pre-disposed composition of soil, leading to the destruction habitat of creatures and organisms. Similarly, dredging of contaminated materials will result in the regrouping of harmful particles and contaminate large of areas of water bodies. Though steps have been taken to mitigate the effects of dredging on the marine environment, several cases involving the destruction of underwater lives are still extensively heard about.

NOx and SOx

Nitrogen oxides (NOx) and Sulfur oxides (SOx) – the two main pollutants found in shipping emissions- has badly affected both marine environment and ozone layer in a number of ways. Both NOx and SOx are combustion products that are emitted into the environment in the form of smoke. It is estimated that in 2005, the water bodies around Europe witnessed 1.7 million tonnes of sulphur dioxide (SO_2) emissions and 2.8 million tonnes nitrogen dioxide emissions from international shipping. And, according to the recent studies, these type of air pollution from shipping accounts around for 50,000 premature deaths per year in Europe. However, strict rules have been formed to reduce their levels in ship's emissions. With IMO revising its standards on the sulphur content of marine fuels, the vessels passing through a Sulphur Emission Control Area (SECA) are not permitted, since 2015, to use fuels with more than 0.1% of sulphur. Similarly, the sulphur limit applicable to all marine fuels used internationally will go from 3.5% to 0.5% since 2020.

Ocean Acidification

The issue of ocean acidification is quickly becoming a threat to both marine lives and humans. Ocean acidification is the continuing decrease of seawater pH caused by the absorption of carbon dioxide (CO_2) from the atmosphere. Ocean acidification has the power to greatly endanger the lives of marine organisms and also humans who depend on fish and fish products for their daily sustenance. Studies have shown that the decreased pH levels affect the behaviour of several marine species, putting them at life-threatening risks.

Sea Water Level Rising

Global warming is alarmingly increasing seawater levels, threatening the marine ecosystem. According to reports, the annual rate of seawater rise during the past two decades has been 0.13 inches a year, which is around twice the average speed of rising over the preceding 80 years. Thus, it is high time that we educate ourselves about the causes and effects of seawater level rising to save the marine environment and biodiversity. The rise in seawater level means more wetland flooding, destructive erosion and agricultural land contamination and more importantly a serious threat to the habitat of several plants, fishes and birds.

Ozone Depleting Substances

Ozone-depleting substances such as CFCs and Halon along with other pollutants from ships are destroying the ozone layer. Ozone Depleting Substances omitted by ships across the world include Methyl Chloroform, Methyl Bromide, Bromochlorodifluoromethane and Bromotrifluoromethane etc. These man-made gases are capable of destroying ozone and in effect, these gases causing harm to the marine environment in several ways.

Waste Pollution from Ships

As we know, tens and thousands of ships are responsible for more than 90 per cent of world trade. Apart from other pollutants such as oil and gas, the waste and garbage generated on board ships poses a significant threat to the marine ecosystem. Both solid and liquid waste in form of ballast water, grey water, food waste, dunnage and packing material, paper products and cleaning material and rags etc. pollutes the seawater and badly affects marine lives. The vessels used for various purposes- be it a container or cruise ship- contribute to this pollution in different levels.

Noise Pollution from Ships

It has been scientifically proved that the noise generated from shipping operations is harmful to marine organisms. Harmful effects of noise pollution on marine organisms include haemorrhages, changed diving pattern, migration to newer places, and damage to internal organs and an overall panic response to foreign sounds. Source of noise pollution from ships include everything from engine noise to the entertainments in cruise ships. The intensity of noise pollution is higher in marine environment since noise travels greater distances easily in water and at the same time, marine life is extremely sensitive to noise due to their heavy reliance on underwater sounds for basic life functions.

Oil Spills

No discussion on marine environment can come to a conclusion without mentioning the biggest cause of marine pollution – oil spills. The world has witnessed several oil spill disasters that have been one of the major concerns of pollution to the marine environment. Disasters such as the Exxon Valdez Oil Spill and Deepwater Horizon etc. have resulted in the extreme pollution of the marine ecosystem, killing thousands of marine species. The oil spilt destroys the insulating ability of several marine species and also the water repellency of bird's feathers, exposing these creatures to life-threatening risks.

Plastic Pollution

It is important to mention this environment degrading agent separately for the sole fact that it has and is the reason for several environmental problems both at sea and land. Those who have read or seen the Pacific garbage patch knows the extent of damage this substance is causing to the marine environment. It is estimated that around 8 million tonnes of plastic waste enters our oceans every year, and by 2050, at this rate, we would witness more plastic than fish in the water bodies across the world. The ill effect of plastic pollution is wide-ranging. The plastic pollution has a direct effect on wildlife as it- plastic bags, fishing nets and other debris-chokes tens and thousands of seabirds and sea turtles every year. The ingestion micro plastics fish and other species also pose risk to their life as well as humans.

The above-mentioned reasons for marine pollution might not be enough to depict the seriousness of the matter. However, it is our humble effort to educate and inform people about the ever-growing threat from these polluting agents.

It's time to educate ourselves. It's time to take some serious action.

Although marine ecosystems provide essential ecosystem services, these systems face various threats.

Human Exploitation and Development

Coastal marine ecosystems experience growing population pressures with nearly 40% of people in the world living within 100 km of the coast. Humans often aggregate near coastal habitats to take advantage of ecosystem services. For example, coastal capture fisheries from mangroves and coral reef habitats are estimated to be worth a minimum of $34 billion per year. Yet, many of these habitats are either marginally protected or not protected. Mangrove area has declined worldwide by more than one-third since 1950, and 60% of the world's coral reefs are now immediately or directly threatened. Human development, aquaculture, and industrialization often lead to the destruction, replacement, or degradation of coastal habitats.

Moving offshore, pelagic marine systems are directly threatened by overfishing. Global fisheries landings peaked in the late 1980s, but are now declining, despite increasing fishing effort. Fish biomass and average trophic level of fisheries landing are decreasing, leading to declines in marine biodiversity. In particular, local extinctions have led to declines in large, long-lived, slow-growing species, and those that have narrow geographic ranges. Biodiversity declines can lead to associated declines in ecosystem services. A long-term study reports the decline of 74–92% of catch per unit effort of sharks in Australian coastline from 1960s to 2010s.

Pollution

- Nutrients,

- Sedimentation,

- Pathogens,

- Toxic substances,

- Trash and microplastics.

Invasive Species

- Global aquarium trade,

- Ballast water transport,

- Aquaculture.

Climate Change

- Warming temperatures,

- Increased frequency/intensity of storms,

- Ocean acidification,

- Sea level rise.

Primary Causes of Destruction of Marine Biological Diversity

Overfishing

Overfishing occurs when fish and other marine species are caught at a rate faster than they can reproduce. We now know without a doubt that the fish in the ocean are a finite resource. Many marine scientists now believe that overfishing is the biggest threat to the ocean environment, even greater than that of other human caused disruptions like increasing pollution. The high demand for fish, along with more effective fishing techniques, has lead to many species of fish around the world being depleted, making them commercially extinct (not worth fishing). As the Ecosystem Overfishing illustration shows (Pew Oceans Commission), some of the modern fishing techniques cause additional unintentional destruction. Between 1950 and 1994, the ocean fishing industry increased the total catch by 400%. In 1997-1998, the total global capture peaked at an estimated 93 million tons. In subsequent years, the total capture has been reduced. This reduction was in part due to climactic changes. However, it is also believed that this is an indication that humans are now fishing more than what the ocean can produce. The growing industry has been increasing production to help meet the growing demand of fish.

World capture fisheries and aquaculture production.

If global overfishing continues, wild fish populations will be further reduced, regardless of how many boats are used or what techniques are employed. Today, most of the world's major commercially valuable fish populations are overfished, and the remainder is exploited at their maximum possible level. (National Oceanic Atmospheric Administration) In 1999, the United Nations Food and Agriculture Organization estimated that 70-78% of worldwide marine fish stocks require urgent intervention to prevent population declines and to help rebuild species depleted by over fishing. In the waters surrounding the United States, the U.S. Department of Commerce listed 959 different fish stocks in the 2001 Annual Report to Congress on the Status of U.S. Fisheries. Of these nearly 1000 species, the status of 68.3% were listed as unknown; of the remaining 31.7%, almost one third are listed as overfished.

The wild fish in our oceans are the last wild creatures that people hunt on a global scale. This overfishing not only depletes the fish that are desirable to consumers but causes serious consequences for the marine environment. The natural ocean ecosystem is disrupted; ultimately threatening many non-fished marine species as their natural food supply is removed. Sea lions, fur seals, and otters, as well as many types of bird are examples of other species that have been placed at risk as a result of overfishing. Removing excessive quantities of specific species has been shown to place the marine ecosystem as a whole at risk of collapse.

One example occurred in the Chesapeake Bay when overfishing and other environmental toxins depleted the oysters. These filterfeeders play an important role in balancing the most abundant ocean plant, microscopic algae. The Chesapeake Bay now has an estimated 1% of the former amount of oysters. The lack of algae caused a disruption of the oxygen balance which has resulted in life-depleted areas known as dead zones. The Chesapeake Bay's "dead zone" now stretches for hundreds of square miles during the summer.

Overfishing impacts not only the natural balance and health of our ocean but has also resulted in great financial loss. The ability of the ocean to produce fish is of vital importance to an estimated 200 million people worldwide as they depend upon the ocean for jobs and for food. According to the United Nations, one in every five humans depends on fish as the primary source of protein. The fishing industry, governments, environmental scientists as well as consumers must all work together to learn how to stop destructive fishing and restore depleted fisheries. Properly managed, our oceans can continue to produce an abundant supply of fish indefinitely. Marine reserves have been one very effective potential solution to the overfishing problem, as the illustration from the PEW Oceans Commission depicts.

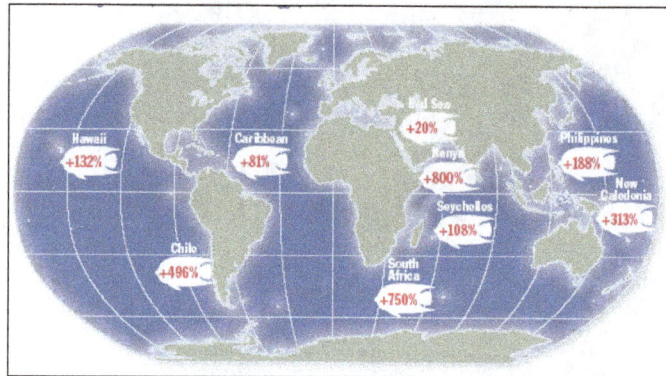

Marine Reserves Increase Fish Biomass.

Around the world, marine reserves have demonstrated the ability to increase fish biomass doubled within five years. The larger fish found within reserves also produce more eggs. For example, ling cod within a reserve in Washington state produced 20 times more eggs per unit area than cod outside the reserve.

Habitat Alteration

Habitat alteration is defined as a change in the particular environment or place where organisms or species tend to live. Habitat alteration is a topic that, by definition, includes many other issues such as pollution, invasive species, overfishing, and aquaculture. However, there are issues that are having a negative impact on our ocean which are not included within these other categories.

Physical damage is one type of environmental damage not included within the other areas. This destruction can be caused by poor fishing gear, cruise ships or even coastal development.

Bottom trawling is a commercially used fishing method where large bag-shaped nets are pulled along the sea floor. This technique is used in search of bottom dwelling species such as shrimp, cod, rockfish and flounder. The nets are used from shallow depths as little as 50 ft (15.2 meters) all the way to depths of 6,000 feet (1829 meters) on the continental shelf. The nets can rise to 40 feet (12.2 meters) tall and be as wide as 200 feet (60.9 meters). This type of fishing was originally used primarily in sandy flat bottom areas in search of fish like flounder. However, with technical innovations such as the use of rollers and rockhopper devices, these same techniques are now used in much more sensitive complex ecosystems such as deep-sea coral forests, rock pinnacles, and boulder-covered areas.

As this fishing gear drags and rolls along and digs into the sea floor, the result is often habitat destruction. Some single rockhopper devices weigh several hundred pounds, each having great potential to cause serious long-term damage to the ecosystem along the seafloor. This type of fishing often sweeps up or kills non-targeted marine life. See our section on for more information on this topic. According to information contained in the PEW Oceans Commission 2003 Report on America's Living Oceans, "typical trawl fisheries in Northern California and New England trawl the same section of sea bottom more than once per year on average. Bottom-dwelling invertebrates can take up to five years or more to recover from one pass of a dredge." This results in sea floor areas that never get a chance to return to their natural state.

The cruise ship industry is known to cause several different types of environmental damage. Within the habitat destruction category, some are accidental such as groundings, which can be devastating. Other causes are, unfortunately, the result of more regular practices and are, hence, avoidable. One issue is their massive anchors, weighing up to 5 tons, which along with the chains, devastate the sea floor which ranges from sensitive coral reefs to sea grass beds. Just one anchoring in calm seas with no wind can do damage that will take a reef 50 years to repair. Government scientists in Grand Cayman report that more than 300 acres of coral reefs have been lost to cruise ship anchors just in the harbor of George Town alone. In the U.S. Virgin Islands National Park, a single anchor drop from a cruise ship in 1988 led to the destruction of almost 300 square meters of reef. Monitoring at this site reveals no significant recovery of hard coral 8 years later. For more information on other types of damage caused by cruise ships.

"A survey of 186 boats in 1987 revealed that 32% were anchored in seagrasses and 14% in coral communities. About 40% of the anchors in coral and 58% in seagrass beds caused damage. Small boats continue to run aground on reefs within Buck Island Reef National Monument and Virgin Islands National Park. The installation of mooring buoys and limits on the size of vessels allowed in park waters have resulted in less pressure on these reefs, but in some areas there is little coral left to protect."

Coastal development is becoming an increasing threat to our oceans as more and more people move closer to the coast. As of present time, 60% of the world's 6 billion people live within 60 miles of a coast. Within the United States, the state of Florida has experienced some of the greatest population growth, increasing from less than 2 million in 1940 to now over 14 million. As a result of development, over half of the Everglades have been lost. In California, according to information published in by The Resources Agency of California "Land reclamation activities, including agriculture and urbanization, have resulted in the loss of more than 90% of the state's historic distribution of riparian and freshwater wetlands." California's Ocean Ecosystem also states that "ninety percent of emergent wetland habitats and more than half the mudflats in the enclosed waters zone have disappeared" and that more than two thirds of the states estuaries and bays have been eliminated. The PEW Oceans Commission reports that "Sprawl development is consuming land at a rate of five or more times the rate of population growth in many coastal areas."

Coastal development threatens our oceans primarily by removing or changing our coastal marshes and estuaries. These areas are of vital importance because they serve as nurseries and spawning areas for aquatic wildlife. These natural areas also filter out pollutants and sediments. They also help to regulate peaking flood levels. Marsh areas in the U.S. alone are disappearing at a rate of 20,000 acres per year. Louisiana has lost half a million acres of wetlands since the 1950's.

The threat to the coastal ecosystem in not always a result of direct development within the estuary or marsh area. In fact, most of the damage occurs as a result of altered natural water flow to these areas which often is a result of the creation of dams, bridges, roads and impervious surfaces such as parking lots. Roads, parking lots, rooftops, and other impervious surfaces typically makeup about 40% of the surface area of suburban development. The impervious surfaces not only alter the natural flow of fresh water to the oceans but they also collect pollutants from automobiles, fields, lawns, and the like which provide a quick transfer to rivers, estuaries, and finally to our oceans. For example, a one-acre parking lot produces about 16 times the volume of runoff that would come from a one-acre meadow. Numerous studies show that when more than 10% of the acreage of a watershed is covered with impervious surfaces, its aquatic biodiversity begins to decline.

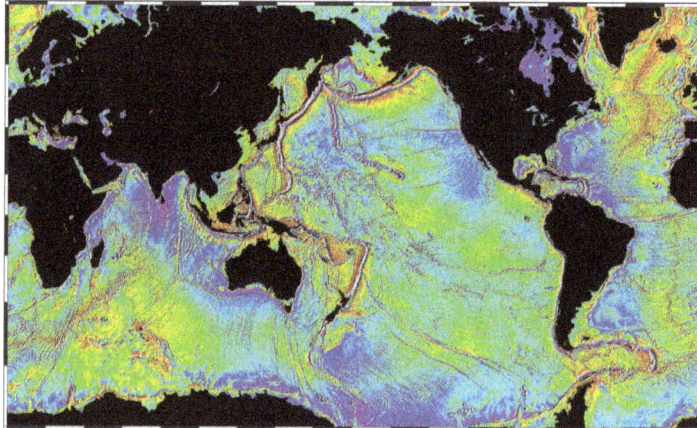

Marine Habitat Destruction

Habitat destruction occurs when the conditions necessary for plants and animals to survive are significantly compromised or eliminated.

Most areas of the world's oceans are experiencing habitat loss. But coastal areas, with their closeness to human population centers, have suffered disproportionately and mainly from manmade stresses. Habitat loss here has far-reaching impacts on the entire ocean's biodiversity. These critical areas, which include estuaries, swamps, marshes, and wetlands, serve as breeding grounds or nurseries for nearly all marine species.

Causes of Ocean Habitat Loss

Humans and Mother Nature share blame in the destruction of ocean habitats, but not equally.

Hurricanes and typhoons, storm surges, tsunamis and the like can cause massive, though usually temporary, disruptions in the life cycles of ocean plants and animals. Human activities, however, are significantly more impactful and persistent.

Wetlands are dredged and filled in to accommodate urban, industrial, and agricultural development. Cities, factories, and farms create waste, pollution, and chemical effluent and runoff that can wreak havoc on reefs, sea grasses, birds, and fish.

Inland dams decrease natural nutrient-rich runoff, cut off fish migration routes, and curb freshwater flow, increasing the salinity of coastal waters. Deforestation far from shore creates erosion, sending silt into shallow waters that can block the sunlight coral reefs need to thrive.

Destructive fishing techniques like bottom trawling, dynamiting, and poisoning destroy habitats near shore as well as in the deep sea.

Tourism brings millions of boaters, snorkelers, and scuba divers into direct contact with fragile wetland and reef ecosystems. Container ships and tankers can damage habitat with their hulls and anchors. Spills of crude oil and other substances kill thousands of birds and fish and leave a toxic environment that can persist for years.

Climate Change

Perhaps the most devastating of all habitat-altering agents, however, is climate change. Scientists are still coming to grips with the consequences that excessive atmospheric carbon dioxide and Earth's rapid warming are having on ecosystems. But there is ample evidence indicating that the oceans are bearing the brunt of these changes.

As Earth's temperature rises, it is primarily the oceans that absorb the extra heat. Even small temperature changes can have far-reaching effects on the life cycles of marine animals from corals to whales.

In addition, warmer temperatures cause excess melting of ice caps and glaciers, raising sea levels and flooding estuaries.

High levels of atmospheric carbon dioxide, caused mainly by the burning of fossil fuels, are absorbed by the oceans, where the gas dissolves into carbonic acid. This elevated acidity inhibits the ability of marine animals, including many plankton organisms, to create shells, disrupting life within the very foundation of the ocean's food web.

Sea Changes

Ongoing efforts to safeguard ocean habitats include the creation of gigantic marine sanctuaries where development is curtailed and fishing is prohibited. Laws banning the dumping of sewage and chemicals into the ocean and policies that foster better stewardship of wetlands are having positive effects. But scientists agree that drastic measures will be needed to avert the ocean crises being created by climate change.

Toxic Pollution

Oil is perhaps the most publicly recognized toxic pollutant. Large tanker accidents like the Exxon Valdez quickly become known worldwide. Events like this, where the Exxon Valdez grounded on

Bligh Reef spilling nearly 11 million gallons of oil into the Prince William Sound in March of 1989, are dramatic and devastating to the entire surrounding marine ecosystem for many years.

The grounding of the Valdez may have been one of the most publicized tanker accidents in recent history, but there are many events like this which have occurred. The biggest spill ever recorded happened during the 1991 Persian Gulf War when about 240 million gallons spilled from oil terminals and tankers off the coast of Saudi Arabia.

Many people don't realize that hundreds of millions of gallons each year quietly end up in our oceans by sources. The following list shows how much oil reaches our oceans from other sources.

Source	Million gallons/year
Large Spill Accidents	37
Routine Ship Maintenance	137
Drains and Runoff	363
From Air Pollution	92
Natural seepage	62
Offshore Drilling	15

These sources result in an estimated average of 706 million gallons of oil pollution entering our oceans each year. Of this, less than 10% is from natural seepage of oil from the ocean floor and eroding of sedimentary rock. The remaining 644 million gallons comes from human activities. Offshore drilling, as a result of accidental spills and other operations, accounts for just over 2%. Large tanker spills, which are reported the world over, account for just over 5%. Air pollution from cars and industry accounts for just over 13% of the total, as the hundreds of tons of hydrocarbons land in our oceans from particle fallout aided by the rain, which washes the particles from the air. Almost 4 times the amount of oil which comes from the large tanker spills, 19%, is regularly released into the ocean from routine maintenance, which includes boat bilge discharge as well as other ship operations. By far, the greatest cause of oil in our oceans comes from drains and urban street runoff. Much of this is from improper disposal of engine oil. An average oil change uses 5 quarts of oil, which alone can contaminate millions of gallons of fresh water. More than half of all Americans change their own oil but only about one-third of the used oil from do-it-yourself oil changes is collected and recycled. In fact, the yearly road runoff from a city of 5 million could contain as much oil as one large tanker spill.

Crude oil from tanker accidents and offshore drilling is most likely to cause problems that are immediately obvious. Most people have seen the images of oil-coated animals and the large oil slicks surrounding the tankers after an accident. As the picture on the right shows (NASA, Jet Propulsion Laboratory) the oil will spread over large areas often continuing to cause harm for many years.

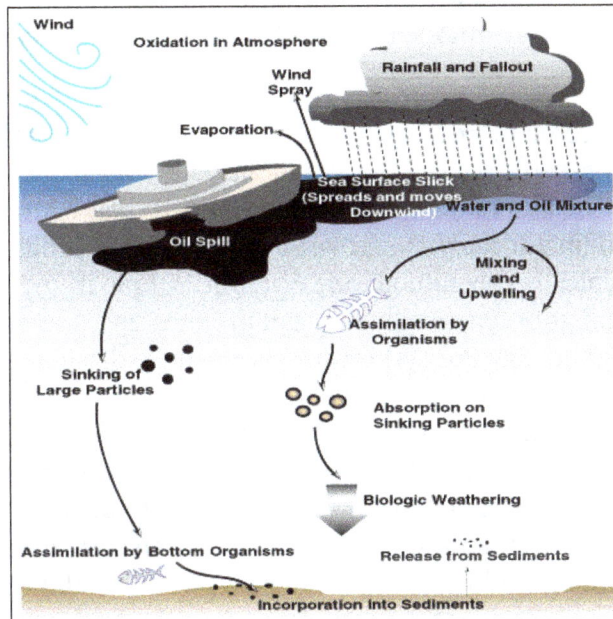

When quantities of surface oil are sufficient to coat animal fur and feathers, the animals cannot stay warm and will ingest the toxic oil while attempting to clean themselves. Many of these oiled animals will freeze to death or die as a result of ingesting these toxins. Many marine animals that do not die quickly as a result of the oil spill may develop liver disease and reproductive and growth problems because of ingestion. Even very small quantities of oil will spread, floating on the surface of the water covering vast areas of water. These thin sheets can kill marine larvae which in turn will reduce the number of marine animals. Effects on human populations are realized through potential health hazards as well as economic losses, such as those associated with the loss of fisheries or tourism. Particularly susceptible to injury from releases of oil are exposed shorelines, shallow reef environments, estuaries, mangrove forests, and wetlands.

Thousands of other pollutants also end up in the ocean. More than 2.8 billion gallons of industrial waste water per day are discharged directly into U.S. ocean waters, excluding electric utilities and offshore oil and gas effluents. Heavy metals released from industry, such as mercury and lead, are often found in marine life, including many of those often consumed by humans. The longer-lived, larger fish such as king mackerel, tilefish, swordfish and shark often contain harmful levels of the pollutant mercury which can harm the developing brain and nervous system of children and fetuses. The list of dangerous chemical pollutants is long, including chemical contaminants like pesticides, pharmaceutical agents, and biological contaminants like bacteria, viruses, and protozoa. Dioxins from the pulp and paper bleaching process can cause genetic chromosomal degradation in marine animals and may even cause cancer in humans. PCB's (polychlorinated biphenyls), which usually come from older electrical equipment, typically cause reproduction problems in most marine organisms. Poly-aromatic hydrocarbons (PAH) are another source of marine toxic pollution and typically come from oil pollution and burning wood and coal. These PAH's are responsible for causing genetic chromosomal aberrations in many marine animals.

Cyanide fishing is a practice still widely used to catch live aquarium fish in the South Pacific and Southeast Asia. Fishermen stun fish by squirting cyanide into the reef areas where these fish seek refuge. They then rip apart the reefs with crowbars to capture disoriented fish. But cyanide is also a killer of coral polyps and the symbiotic algae and other small organisms necessary for healthy oceans.

Many cruise ships have become more like floating cities, transporting millions of people into the most pristine ocean environments around the world. They have been operating with little to no environmental regulations. The lack of regulation by this industry has caused a great deal of damage to sensitive marine environments and is of growing concern as the industry is rapidly expanding. Ships have grown from typically accommodating 600-700 people in 1970 to ships that now carry over 5000 people. A typical 3000 passenger ship can produce 255,000 gallons of wastewater and 30,000 gallons of sewage every day. All of this waste is normally discharged directly into the ocean and it is legal to do so in most areas provided the boats discharge 3 nautical miles from the coast. This waste can contain bacteria, pathogens, medical waste, oils, detergents, cleaners, heavy metals, harmful nutrients (nitrogen amongst others) and other substances. These substances can be brought back to coastal areas as well as cause serious damage to the aquatic life further out in the sea, including posing a risk for contaminating seafood. Nitrogen compounds can also contribute to environmentally hazardous algae blooms. Typically 75-85% of the solid waste from a ship is incinerated at sea adding to sea pollution as the toxins and ash settles back into the ocean. Additionally, habitat alteration is a common occurrence where cruise ships use anchoring systems. The sheer size and weight of an anchor dramatically disturbs the seafloor.

Some cruise lines have now worked to introduce and use more environmentally responsible methods. However, there remains a lack of laws and those laws that do exist, are inadequately enforced.

Another serious type of marine pollution is nutrient pollution. This pollution is caused primarily from agricultural runoff that contains fertilizers and growth stimulants as well as from airborne nitrogen compounds that comes from automobile exhaust, industrial pollution and ammonia from manure. This has long been considered a problem in freshwater systems. In recent years, scientists have become more concerned about eutrophic (overly nutrient enriched) conditions in coastal estuaries. Nutrient pollution now represents the most widespread pollution problem facing U.S. coastal waters. Nutrient pollution causes many problems, including:

- Harmful algal blooms.

- Hypoxic areas or "dead zones".

- Fish kills.

- Loss of seagrass and kelp beds.

- Coral reef destruction.

- Eutrophication.

Eutrophication is a condition in an aquatic ecosystem where high nutrient concentrations stimulate blooms of algae (e.g., phytoplankton). The main cause of eutrophication is excess nitrogen run-off from farm fertilizers, sewage and industrial pollutants. Eutrophication reduces water clarity and depletes oxygen. Reduced water clarity can starve sea grasses and algae that live in corals from light, reducing their growth or killing them. While wind and waves aerate surface waters, the pycnocline—a layer of rapid change in water temperature and density—acts as a barrier to oxygen exchange in bottom waters.

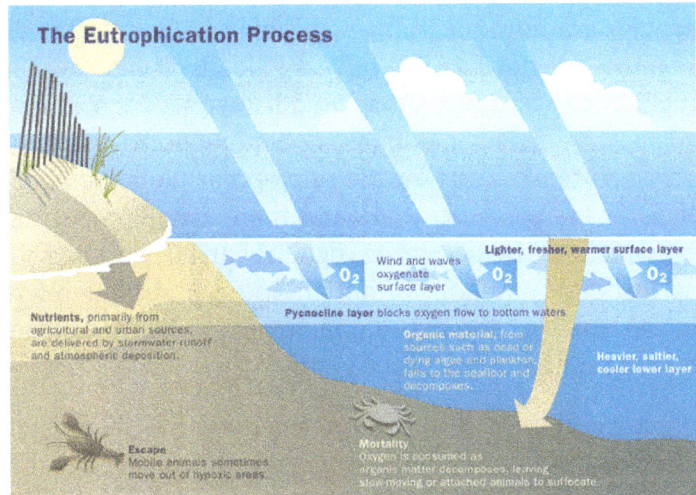

Excess phytoplankton reduces water clarity and consumes oxygen. Phytoplankton need nutrients as well as the energy from the sun to survive but too many nutrients can cause algae blooms and, in turn, red tides (dying phytoplankton). In some regions (particularly near major rivers), excess nutrients can be added to the coastal zone as a result of fertilizer runoff, sewage, animal feedlot runoff, or air pollution. During the bloom, the phytoplankton consume nutrients and oxygen which, in turn, causes a decrease in the amount of dissolved nitrogen and phosphorus in the water body. As the nutrients become depleted, the algae can no longer survive. The dead phytoplankton sink to the bottom of the water column where they are consumed by decomposers. Since these decomposers require oxygen to break down the algae, dissolved oxygen levels will decrease during this time period. Resulting low oxygen levels can be detrimental to fish health; if dissolved oxygen drops to below 2 mg/l, mass fish kills can result. This is known as hypoxia. The areas in which hypoxia has occurred are known as 'Dead Zones.' Dead zones have been a factor in the Gulf of Mexico and Chesapeake Bay on the U.S. east coast, and are now spreading to other bodies of water, including the Baltic Sea, Black Sea, Adriatic Sea, Gulf of Thailand and Yellow Sea. There are now nearly 150 dead zones around the globe-- double the number in 1990, with some extending 27,000 square miles. The article states that "Unless urgent action is taken to tackle the sources of the problem, it is likely to escalate rapidly."

Invasive Species

Invasive species is the term used to describe plants and animals introduced to new areas where these same species do not belong. The introduction of invasive species has been a problem for as long as ships have been traveling the seas. Both plants and animals from land and sea are moved in, on and under ships from their existing habitat to new areas where they can become an invasive species. When these foreign species are introduced into a new habitat and environmental conditions are favorable, these non-indigenous species become established. They, in turn, can compete with or prey upon native species of plants, fish, and wildlife. They may also carry diseases or parasites that affect native species because they have no resistance. This can disrupt the aquatic environment and economy of affected near-shore areas.

Most of the invasive species are introduced from shipping ballast tanks which provides balance and weight to ships when they are not carrying full cargo loads. It can contain a wide variety of microscopic marine life including eggs, cysts, larvae, and bacteria. Global shipping now moves about 80 percent of the worlds' commodities, making this phenomenon a grave threat to all marine environments. "Every assessment indicates that the rate of marine introductions in U.S. waters has increased exponentially over the past 200 years and there are no signs that these introductions are leveling off. New introductions are occurring regularly on all coasts, producing immediate and damaging impacts, and leading to millions of dollars in expenditures for research, control, and management efforts. In the San Francisco Bay alone, for example, an average of one new introduction was established every 14 weeks between 1961 and 1995".

It is estimated that there are hundreds of invasive species within U.S. coastal waters and that they may cost the U.S. hundreds of millions of dollars each year. On average, some two million gallons of ballast are released into U.S. waters alone each hour, resulting in the introduction of potential invasive species. "It has been estimated that exotic species that become invasive cost the United States $137 billion annually more than earthquakes, floods and fires combined."

Shipping is not the only way in which invasive species are introduced. Marine debris can also allow non-native species to travel great distances where they can eventually be introduced into a new area. The slow rate of travel associated with slower environmental changes can allow the transplanted plants and animals to have a high survival rate. In a survey of some 30 remote islands, David Barnes of the British Antarctic Survey in Cambridge, UK, found that human debris and litter had more than doubled as a vehicle for marine organisms to transport themselves to remote lands. Barnes is further quoted as saying "We estimate that rubbish of human origin in the sea has roughly doubled the propagation of fauna in the subtropics and more than tripled it at high (> 50°) latitudes, increasing the potential for alien invasion and adding to the problems already created by sea-borne plastic materials in the form of injuries and mortality among marine mammals and birds." A more recent cause of invasive species problem has been practices within the Aquaculture industry.

The Aquaculture Industry

During the last generation as the human population increased, our demand for fish has grown tremendously. The commercial fishing industry has been able to meet this demand by creating better fishing techniques. We have now reached a point when the demand is so high and the technique so effective that many ocean areas have been over-fished and are now depleted of the fish quantities that are sufficient to meet the growing demand.

One potential solution to this problem is aquaculture. Aquaculture is quite simply the production of fish and other aquatic species by using farming techniques. This method of supplying fish has, in recent years, been growing at a tremendous pace. "Aquaculture is now one of the fastest growing sources of protein, expanding at 10 percent per year. Output more than tripled between 1984 (the first year global aquaculture statistics were compiled by FAO) and 1996, from 7 million tons worth $10 billion, to 23 million tons valued at $36 billion. Today, one out of every five fish consumed comes from the farm."

While this could be a sustainable method of meeting the seafood demands of the exploding human population, it to date has shown to have some serious problems;

- Farmed Fish Escape:
  - There have been many large scale escapes introducing hundreds of thousands of farm-raised fish into the wild. In one single incident in 2000, the population of salmon in Maine was elevated by 1000 times when approximately 100,000 fish escaped. Contrary to what some people may think this is not beneficial to the marine environment, as it does not "just replace the natural number of fish".

- The farmed fish are often not native to the areas in which they are farmed. This practice often displaces the native species.

- Many fish farms use antibiotics and genetically altered fish, known as transgenic fish. They are altered to grow fast to a large size while consuming less food. They often are not permitted to even live to sexual maturity.

- Farmed fish may have diseases not found in the wild. The wild fish have no natural resistance to these diseases.

- There have been studies which show that farm raised fish are more aggressive breeders than their wild counterparts.

- Farmed fish now outnumber wild fish 48 to 1 in the North Atlantic.

- Aquaculture alters the natural ecosystem:
  - Thousands of acres of mangroves are altered to make room for shrimp farms, displacing those species which naturally exist there, thus forever altering the ecologically important mangrove areas.
  - To provide food for the farmed fish, trawlers vacuum the sea for anchovies and mackerel to make fish meal, effectively removing these fish from the ocean's food chain.
  - The genetically altered farmed fish pose a risk to the genetic quality of wild fish, as escaped farmed fish are now interbreeding with wild fish.
  - Rotting, uneaten food pellets from fish pens pollute surrounding areas. These pellets, combined with the concentrated excrement from the farmed fish, often contain antibiotics and dioxins which contribute to ecosystem destruction and, in some cases, "Dead Zones".
  - Sea lice infestations from high concentrations of farmed salmon have caused health problems to wild salmon.

- The farmed fish can be a serious health hazard to human consumers:
  - "Analysis of Fish Consumption Data Shows 800,000 U.S. Adults Eat Enough PCBs From Farmed Salmon to Exceed Allowable Lifetime Cancer Risk 100 Times Over". "The

Environmental Working Group bought the salmon from local grocery stores and found seven of 10 fish were so contaminated with PCBs that they raise cancer-risk concerns, relative to health standards of the US Environmental Protection Agency (EPA)". The toxin levels are believed to be higher as farmed fish are often fed ground-up junk fish (fish with little other commercial value) captured close to shore where these toxins levels are higher.

- "There's no doubt that fish is good for you, but these data suggest if you're going to eat farmed salmon then eating it a little less frequently would be a good idea."

- May be a waste of resources:

  - To produce one pound of farmed salmon, 2.4 to 4 pounds of wild sardines, anchovies, mackerel, herring and other fish must be ground up to render the oil and meal that is compressed into pellets of salmon chow. Resulting in a 58 percent to 75 percent loss of protein and food energy.

What can you do about it?

- If you are a boater, do not carry your ballast or bilge water into new environments.

- After boating, carefully clean the boat's hull and propellers before introducing the boat to a new area.

- If you fish, dump your bait buckets in the trash, not local waterways.

- Do not release aquarium contents into local waterways.

- Choose seafood safe for your family and the environment. This is one issue over which consumers have the power make change. There is a lot of information available on the web to guide you. One site we find easy to use is by The Monterey Bay Aquarium. They provide a downloadable/printable wallet-sized card to help you buy the seafood that is healthy for consumption and the environment as well as more information about the issues surrounding sustainable fishing methods.

Driving Forces and Pressures

Driving forces are factors that lead to environmental change. Pressures on the environment stem from these driving forces, resulting in changes to the quality or condition of the ecosystem. The primary driving force behind marine introductions is the transportation, trade practices, and other activities of humans. Transport mechanisms, or vectors, associated with human activity include transport of species within or on ships, aquaculture, intentional introductions, and trade (pet trade, live seafood). In addition, other pressures such as habitat modification and climate change alter the survival rates of both native and non-native species and may influence establishment of marine invaders.

Global Trade and Exploration

Humans have introduced marine species to new environments inadvertently, and at times purposely, for centuries. The rise in global trade through commercial shipping in particular has dissolved historical barriers for distribution of marine organisms and has led to an unprecedented increase in the rate of marine introductions in the last 200 years. In the Gulf of Maine, the majority of marine invaders originate from Europe.

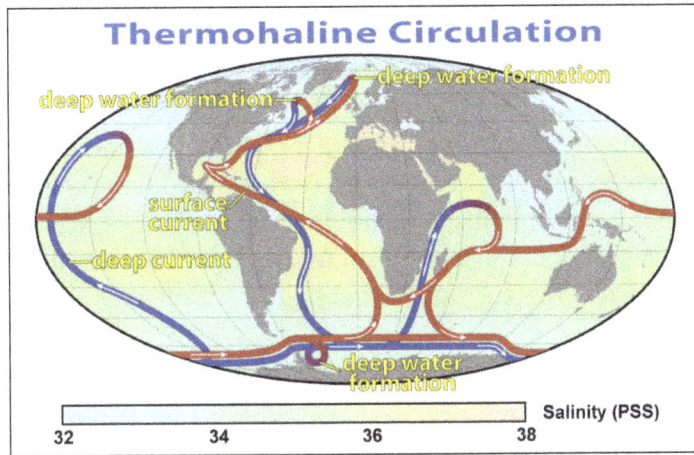

Origin of marine invasive species in the Gulf of Maine.

highlighting the importance of global trading routes between the northeast and northwest Atlantic in marine introductions. Commercial shipping has led to the introduction of marine invasive species into the Gulf of Maine through two primary mechanisms: transport by ballast and fouling on ship surfaces (hull, sea chest, etc.).

Table: Species introduced by shipping in the Gulf of Maine.

Species	Vector
Furcellaria lumbricalis (red alage) B	Ballast Water
Ovatella mysotis (mouse ear snail)	Rock Ballast
Fucus serratus (brown algae)	Rock Ballast
Littorina littorea (common periwinkle)	Rock Ballast
Neosiphonia harveyi (red algae)	Fouling
Porphyra katadae (nori, red algae)	Fouling
Botryllus schlosseri (star tunicate)	Fouling
Diplosoma listerianum (tunicate)	Fouling
Codium fragile ssp fragile (green fleece)	Fouling
Antithamnion pectinatum (red algae)	Shipping, unspecified
Bonnemaisonia hamifera (red algae)	Shipping, unspecified
Lomentaria clavellosa (red algae)	Shipping, unspecified
Melanosiphon intestinalis (brown algae)	Shipping, unspecified
Convoluta convoluta (flatworm)	Shipping, unspecified

Ballast

The use of solid ballast obtained from intertidal habitats of Europe may have transported entire communities to the Gulf of Maine, and has been implicated in the introduction and subsequent spread of the brown algae Fucus serratus and snail Littorina littorea to Nova Scotia in the 18th Century. In the last half century, an increased number of commercial vessels, reduction of toxins in ballast water, and larger capacity of ballast tanks have improved the survival of marine invaders in transit and thus the number of viable marine introductions. Mysids, amphipods, cladocerans, copepods, numerous microscopic planktonic organisms, algal filaments, and fish have been observed to survive in ballast tanks in journeys lasting nearly two weeks, while polychaete larvae and copepods can survive voyages of 30 days or more. In the Gulf of Maine, it is hypothesized that the red alga Furcellaria lumbricalis was introduced via ballast water.

Fouling

Fouling is the accumulation of marine organisms on the hull, sea chest, and other surfaces of ships. Distinguishing between introductions resulting from fouling versus ballast water is extremely difficult. However, species that have short-lived larvae, which are not likely to survive long journeys in ballast

water, are candidates for introductions through fouling. For example, in the Gulf of Maine it is hypothesized that the colonial tunicates Botrylloides violaceus and Diplosoma listerianum were fouling introductions due, in part, to their short larval period. Modern vessels are faster, have shorter times in port, and are more frequently maintained, thus the role of hull fouling in recent transoceanic introductions has been questioned. However, Drake and Lodge recently collected close to 1,000 live organisms from the hull of a single cargo ship entering the Great Lakes from Algeria, indicating that the importance of this vector should not be overlooked. There are at least five species in the Gulf of Maine that may have been introduced by fouling.

Jim Frazier.

Pressures from other Human Activities

The high concentration of people living on the coast and their associated economic and social activities results in numerous pathways (vectors) that can facilitate the regional spread of marine introduced species. For example, transport of species for aquaculture is implicated in several marine introductions in the Gulf of Maine. Recreational boating and other related activities transport marine invaders from their point of introduction to additional areas, while anthropogenic disturbances to coastal habitats, such as pollution, habitat modification, and climate change may alter survival rates and interactions of native and non-native species.

Aquaculture

Transfer of non-native species for aquaculture, particularly oysters, has been identified as a major vector of marine introductions in North America. Species have been introduced directly to the Gulf of Maine for aquaculture purposes, as with the European oyster, Ostrea edulis, or may be secondarily associated with aquaculture organisms. It is hypothesized that introductions of ubiquitous and aggressive species such as the colonial tunicates Didemnum vexillum and Botrylloides violaceus and the green algae Codium fragile ssp. fragile resulted from the transfer of oysters for aquaculture. In nearby Prince Edward Island, regional transport of invasive tunicates stemming from the movement of bivalves for aquaculture is a known problem. Past import of Anguilla rostrata (American eel) for aquaculture led to the spread of the nematode parasite Anguillicola crassus throughout the Gulf of Maine.

DFO.

Habitat Modification

The role of habitat modification on the introduction and survival of non-native marine species in the Gulf of Maine is not clear. However, increased numbers of non-native species are often seen in areas disturbed by human activities, and successful invaders may possess traits that enable them to perform better in altered habitats relative to native species. It is thought that native species compete best on surfaces for which they are evolutionarily adapted, giving non-native species a competitive advantage on newer, artificial substrates. For example, marine invasive species are often more likely to be present on floating pontoons and pilings than adjacent natural habitat.

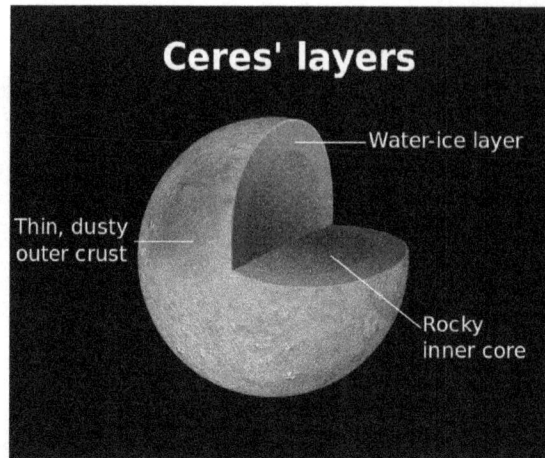

N. Houlihan.

Water quality conditions may also play a role in the establishment of marine invasive species. For example, the red alga Grateloupia tururturu, a recent invader to the Gulf of Maine, is highly tolerant of polluted waters. In eutrophic (nutrient rich) systems, invasive species that are better competitors at high nutrient levels and low oxygen conditions may have an advantage over native species. However, direct relationships between survival of marine invaders and water quality are not always clear; in southern New England both native and non-native ascidians are most diverse in areas of fair water quality and moderate levels of nitrogen. Ironically, recent improvements to water quality, both in local harbors and distant source ports, have been implicated in increased survival of marine invaders both in ballast and at the point of introduction.

Climate Change

Climate change and the resultant modifications in habitat may impact the survival and establishment of species in various ways. Hellmann et al. propose several consequences of climate change on marine invasive species relevant to the Gulf of Maine: altered patterns of human transport (longer shipping seasons, new routes, etc.), altered climatic restraints favoring non-natives or increasing the possibility of survival for previously unsuccessful invasions, altered distributions (range shifts, etc.), and altered impacts. Long-term studies of rocky shores in California have shown latitudinal shifts in species abundance and geographic range boundaries as a result of temperature increases, and similar changes may occur in the Gulf of Maine if temperatures continue to rise. Since organisms are generally most abundant in the center of their range, species with more southerly borders should expand, while those with northern boundaries should decrease. For example, a warming trend during the last mid century is implicated in the expansion of the green crab, Carcinus maenas, from waters south of Cape Cod Massachusetts into the Gulf of Maine. Changes in climate resulting in warmer winter water temperatures in particular could provide a thermal refuge for invading species and may have important impacts to timing of recruitment and survival. For example, it is hypothesized that the degree to which the invasive tunicate Didemnum vexillum degrades in cold weather influences its ability to regenerate and reproduce sexually, thus an increase in winter water temperatures may enhance the reproductive capability of this species. The non-native tunicates Ascidiella aspersa, Botrylloides violaceus, and Diplosoma listerianum recruit earlier in warmer years, and total annual recruitment is positively correlated with mean winter water temperature. The invasive red algae Grateloupia turuturu produces larger blades during warm temperature events, resulting in increased cover. An increase in winter water temperatures has also been linked to a rise in Perkinsus marinus (Dermo disease) outbreaks in oysters .

Regional Transport for Trade and Recreation

Transport of marine introduced species from their point of introduction to other areas regionally can occur through many vectors. Coastwide trade (short-sea shipping) and recreational boating are likely important transport vectors throughout and between the Gulf of Maine, southern Atlantic, and northern Canada. Boat hulls, propellers, chains, anchors, and ropes are easily fouled by marine invaders, facilitating spread when the vessel relocates or is cleaned, particularly for species capable of reproducing through fragmentation, such as colonial tunicates and algae. Domestic arrivals account for about 35 percent of the total ballast water discharged in New England, and ballast water exchange zones are located within Gulf of Maine waters for ship traffic to and from Canada. It is likely that regional transport vectors facilitated the spread of Littorina littorea from the Gulf of St. Lawrence to the Gulf of Maine. Perkinsus marinus, the protozoan resulting in Dermo disease of oysters, and the barnacles Chthamalus fragilis and Balanus subalbidus all have expanded their range northward to the Gulf of Maine from south Atlantic waters, likely aided by domestic shipping vectors.

Impacts

The definition of an invasive species is a non-native or cryptogenic species introduced by humans that causes harm to ecosystem or economic resources. However, deciphering what this harm may be for a particular species is difficult, and there are relatively few empirical studies available on the impacts of invasive species. In general, historical records and pre-invasion data for most species and coastal communities are lacking, thus when an invasion occurs there is no baseline condition from which to evaluate impacts. Although difficult to discern, there are in general three broad categories of impacts from introduced marine species: ecosystem impacts, economic impacts, and human health impacts.

Ecosystem Impacts

Parker et al. developed a model where the impacts of introduced species are measured at multiple levels: effects on individuals, genetic effects, population dynamic effects (community effects), and effects on ecosystem processes. These impacts can work separately or in combination for any species or suite of species and can range from small localized impacts to larger-scale regional impacts. table describes examples of impacts of introduced species on native species in the Gulf of Maine.

One of the most well studied impacts in the Gulf of Maine is community shifts resulting from the cumulative effect of two marine invaders—the bryozoan Membranipora membranacea and the green alga Codium fragile ssp. fragile—on native laminarian kelps. Kelp beds in the Gulf of Maine provide critical habitat for a wide range of species, such as native fish and invertebrates. Historically, grazing by sea urchins was the major source of disturbance in kelp beds, leading to bare patches and, at times, large-scale removal (urchin barrens). In the past, kelp would generally re-establish after most disturbances. However, in the late 1980s, a dramatic transformation of kelp beds began in the Gulf of Maine concurrent with the arrival of M. membranacea.

Examples of marine invasive species impacts on native species in the Gulf of Maine.

Native species	Impact
Mytilus edulis (blue mussel)	• Hemigrapsus sanguineus feeds on juveniles, consumes up to 150 mussels per crab per day in the laboratory, comprises 30% of the diet in the field. • Flatworm Convoluta convoluta feeds on juveniles. • Makes up to 45% of the diet of Carcinus maenas.
Mya arenaria (soft shell clam)	• In caging experiments, C. maenas removed 80% of small M. arenaria and consumed up to 22 clams per crab per day.
Homarus americanus (lobster)	• C. maenas arrives to food faster and defends food resources from juvenile lobsters in the laboratory.
Littorina saxatilis (periwinkle)	• Growth rate is reduced when competing with Littorina littorea. • Susceptible to predation by C. maenas.
Ilyanassa obsoleta (mud snail)	• L. littorea competitively displaces I. obsoleta from habitat. • C. maenas and L. littorea feed on egg capsules.
Fucus spp. (rockweed)	• L. littorea can prevent establishment of Fucus on smooth surfaces by grazing small germlings.

Membranipora membranacea, a native of Europe, was first recorded in the Gulf of Maine in 1987, and within three years became the dominant kelp epiphyte in offshore locations of New Hampshire and Maine. Kelp blades encrusted with M. membranacea are more susceptible to breakage and are also heavily consumed by sea urchins and the native gastropod Lacuna vincta. Increased drag and heavy grazing pressure on blades encrusted with M. membranacea leads to extensive kelp defoliation during winter storm events. The resulting large, bare patches within the kelp bed provide space for another invader, the green alga Codium fragile ssp. fragile to establish.

Codium fragile ssp. fragile was introduced to the Gulf of Maine as a hitchhiker on oysters transplanted from Long Island Sound and also may have drifted through the Cape Cod Canal from southeast Massachusetts. This species (further referred to as C. fragile) cannot directly displace standing kelp but opportunistically colonizes areas after disturbance. The pattern of defoliation of kelp blades encrusted with M. membranacea and subsequent colonization of the former kelp bed by C. fragile has occurred to such an extent in the Gulf of Maine that it is now the dominant canopy species in some locations. For example, at the Isle of Shoals, C. fragile increased 20-fold and kelp cover decreased from 44 percent to 2 percent in ten years. Similarly, sites in New Hampshire and Maine have experienced 27-fold increases of C. fragile over a 22-year period.

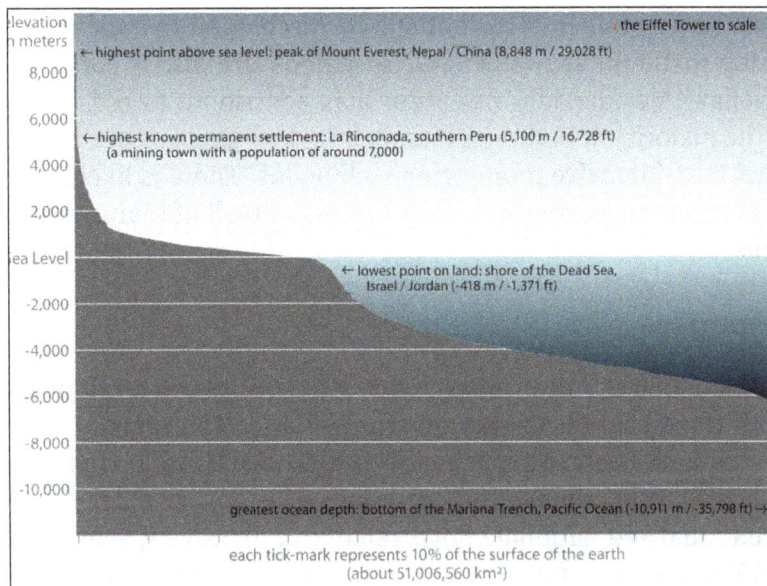

elevation in meters

8,000 — highest point above sea level: peak of Mount Everest, Nepal / China (8,848 m / 29,028 ft)

the Eiffel Tower to scale

6,000

— highest known permanent settlement: La Rinconada, southern Peru (5,100 m / 16,728 ft)
(a mining town with a population of around 7,000)

4,000

2,000

Sea Level

— lowest point on land: shore of the Dead Sea,
Israel / Jordan (-418 m / -1,371 ft)

-2,000

-4,000

-6,000

-8,000

-10,000

greatest ocean depth: bottom of the Mariana Trench, Pacific Ocean (-10,911 m / -35,798 ft) →

each tick-mark represents 10% of the surface of the earth
(about 51,006,560 km²)

MIT Sea Grant.

It is unclear how the transformation of kelp beds to C. fragile meadows impacts ecosystems of the Gulf of Maine. Codium fragile ssp. fragile and kelp morphologies are not similar, and establishment of C. fragile in former kelp beds may modify the habitat to such an extent that there are wholesale shifts in the species assemblage. For example, the native wrasse Tautogolabrus adspersus (cunner) recruits six times higher in kelp than in C. fragile meadows. The rise in C. fragile has also led to an increase in its associated epiphyte, Neosiphonia harveyi, whose population has increased six-fold since 1966. Neosiphonia harveyi is now one of the most widely distributed algal species in the Gulf of Maine.

Economic Impacts

There are few empirical studies focused on the economic impacts of marine introduced species in the Gulf of Maine and elsewhere. In general, economic impacts of non-native species may include monetary costs for management, cost and damages incurred due to fouling of equipment and vessels, aesthetic and/or recreation impacts, and actual losses relative to impacts to fishery or aquaculture resources. In the Gulf of Maine, fouling, particularly by invasive tunicates, and impacts to commercially harvested species is a concern. Fouling by tunicates on gear and harvested product is a major issue for the mussel farming industry in Canada, where approximately 2-3 tonnes of bivalves are harvested each year in New Brunswick and Nova Scotia. Predation by Carcinus maenas on Mya arenia (soft shell clam) is implicated in the decline of harvest and value. Total costs associated with the C. maenas invasion in the US are estimated at $44 million, but it is unclear how this figure was derived. In addition, non-native marine species may result in aesthetic impacts that alter recreation and are costly to clean up. For example, C. fragile often washes ashore and forms large clumps on beaches that are unsightly and result in noxious odors.

Public Health Impacts

Impacts to public health are included in the suite of potential impacts of marine invaders, although not much is known regarding public health impacts in the Gulf of Maine specifically. Introduced pathogens, such as Vibrio cholera, the bacteria responsible for cholera in humans, have been found in ballast water and could potentially be discharged to local waters. Organisms that result in concentrated toxins in seawater and/or seafood are also a concern. For example, the non-native dinoflagellate Alexandrium minutum may contribute to red tide outbreaks south of the Gulf of Maine. The majority of pathogenic and/or toxic organisms are microscopic and require specific, and at times, cost-intensive monitoring techniques. There is likely a wide array of potential invasive pathogens that are currently overlooked in the Gulf of Maine due to a lack of targeted monitoring programs.

Actions and Responses

Focused efforts on the management of marine invasive species are a relatively recent phenomenon. In contrast with terrestrial invasions, government and other agents have been slow to recognize marine introductions as an issue, primarily due to a lack of information and demonstrable impacts to human health, ecosystems, and economies. Responses exist on several scales; there are numerous international and national efforts relative to invasive species in addition to regional and local initiatives. While some efforts include a regulatory component, the bulk of management approaches to marine invasive species are voluntary.

National

In the Gulf of Maine, the primary instrument for prevention of marine introductions is the regulation of ballast water by the US Coast Guard under the National Invasive Species Act and Transport Canada under the Canada Shipping Act. Mandatory ballast water exchange and reporting requirements have been in place for all US waters since 2004 and in Canada since 2000. There has been a recent push by the US Coast Guard to implement ballast water discharge standards similar to those recently proposed by the International Maritime Organization. The process includes a phased approach of performance standards based upon number of living organisms, to be reached primarily through ballast water treatment technologies and an eventual phase out of ballast water exchange. In addition, a recent US Supreme Court decision has led the US Environmental Protection Agency (US EPA) to regulate discharges incidental to the normal operation of vessels, including aquatic invasive species (AIS) in ballast water and biofouling, as a pollutant under the National Pollutant Discharge Elimination System (NPDES). Lodge et al. make several recommendations to improve introduced species management in the United States, including: better management of invasive species pathways, adoption of more quantitative risk assessment procedures, use of cost-effective diagnostic and surveillance activities to improve communication, creation of new legal authorities and emergency funding for rapid response, increased funding to slow the spread of invaders and protect habitats and infrastructure, and the establishment of a national center for invasive species management.

Regional and Transboundary

No official transboundary regulatory effort exists between the governments of the United States and Canada for the Gulf of Maine specifically. The Northeast Aquatic Nuisance Species Panel serves as the major coordinating body for the Northeast region, and all Gulf of Maine states and provinces have delegates that sit on the panel.

Examples of international and national responses to marine introduced species.

International	
United Nations	• Environmental Program, Agenda 21, addresses the issue of aquatic invasives in the context of ballast water and aquaculture. • Food and Agriculture, Code of Conduct for Responsible Fisheries: covers fishing practices and aquaculture. • Convention on Biological Diversity, Article 8(h): commitment to prevent the introduction, and to control and eradicate alien species. • International Maritime Organization: International Convention for the Control and Management of Ships' Ballast Water and Sediments.
International Council for Exploration of the Seas (ICES)	• Code and Practice on the Introduction and Transfer of Marine Organisms : aquaculture focused.

National	
United States	Lacey Act: limited to controlling intentional introductions of injurious species: • Nonindigenous Aquatic Nuisance Species Prevention and Control Act (NANPCA): regulate ballast water in the Great Lakes and establishment of the Aquatic Nuisance Species (ANS) Task Force. • ANS Task Force: coordinates federal activities, provides funding and direction to regional panels, directs states to develop management plans, provides limited funding for plan implementation. • National Invasive Species Act: amended NANPCA to broaden ballast water requirements to the entire US. • Presidential Executive Order No. 13112: created the National Invasive Species Council. • National Invasive Species Council, National Invasive Species Management Plan: serves as national blueprint for invasive species management.
Canada	Canada Shipping Act, Section 657.1: provides for the power to pass ballast water regulations: • An Invasive Alien Species Strategy for Canada: national strategy for invasive species management. • The Invasive Alien Species Partnership Program: provides funding in support of the goals of the Invasive Alien Species Strategy. • Canadian Biodiversity Strategy: provides means to anticipate, identify and monitor alien organisms, screening standards, and risk assessment. • National Wildlife Policy: nonindigenous species should not be introduced into natural systems. • Canadian Environmental Protection Act: applies an ecological risk assessment process before permitting the introduction of any new species • Fisheries Act: develops a standard ecological risk assessment process, specifically in the context of fish stocking, live bait, and aquaculture. • National Code on Introductions and Transfers of Aquatic Organisms: assesses proposals for moving aquatic organisms between water bodies.

In Atlantic Canada, the Canadian Council of Fisheries and Aquaculture Ministers, Aquatic Invasive Species Task Group is responsible for the development of an action plan to address the threat of aquatic invasive species, and the Council of Atlantic Premiers recently signed a Memorandum of Understanding (MOU) for the Development of the Aquaculture Sector that also addresses invasive species. The Marine Monitoring and Information Collaborative, coordinated out of the Massachusetts Office of Coastal Zone Management (MA CZM), is a volunteer early detection and monitoring network for marine invasive species that includes numerous sites in Maine, New Hampshire, and Massachusetts. The Rapid Assessment Survey, a partnership between the Massachusetts Institute of Technology Sea Grant Program, MA CZM, and National Estuary Programs in New England, is conducted by a team of expert taxonomists who examine marine fouling communities in New England and Long Island Sound. The surveys have been conducted roughly every three years since 2000, with the next survey planned for summer 2010.

State and Provincial

Provincial and state management of marine invasive species is largely vector based, with aquaculture as a prime example. States and provinces largely have permitting jurisdiction over operation of aquaculture facilities, including the transfer and source of aquaculture species, and other projects in the marine realm that could potentially introduce or spread marine invasive species. For example, in New Brunswick a permit can be refused if the regulator determines that the proposed operation poses an unacceptable risk to the environment, however the threat of invasive species is not specifically identified. Most state and provinces within the Gulf of Maine do not have programs dedicated to marine invasive species management; rather management often falls under a variety of environmental agencies and programs (Doelle 2001). For example, while Massachusetts has an Aquatic Invasive Species Program, a Management Plan, and an Aquatic Invasive Species Group, these do not hold any regulatory authority. In addition, there are at least six other entities that are involved with aquatic invasive species management to some degree). A tool that states and provinces can use for marine invasive species management is to add on to pending federal regulation in the form of comments or conditions. For example, Maine and Massachusetts added conditions to the recent US EPA Vessel General Permit limiting underwater husbandry, and Massachusetts also included provisions requiring ballast water exchange for ships engaged in coastwide trade.

Marine Conservation

Marine biodiversity is higher in benthic rather than pelagic systems, and in coasts rather than the open ocean since there is a greater range of habitats near the coast. The highest species diversity occurs in the Indonesian archipelago and decreases radially from there. The terrestrial pattern of increasing diversity from poles to tropics occurs from the Arctic to the tropics but does not seem to occur in the southern hemisphere where diversity is high at high latitudes. Losses of marine diversity are highest in coastal areas largely as a result of conflicting uses of coastal habitats. The best way to conserve marine diversity is to conserve habitat and landscape diversity in the coastal area. Marine protected areas are only a part of the conservation strategy needed. It is suggested that a framework for coastal conservation is integrated coastal area management where one of the primary goals is sustainable use of coastal biodiversity.

Marine conservation, also known as ocean conservation, refers to the study of Marine plants and animal resources and ecosystem functions. It is the protection and preservation of ecosystems in oceans and seas through planned management in order to prevent the exploitation of these resources. Marine conservation is driven by the manifested negative effects being seen in our environment such as species loss, habitat degradation and changes in ecosystem functions and focuses on limiting human-caused damage to marine ecosystems, restoring damaged marine ecosystems, and preserving vulnerable species and ecosystems of the marine life. Marine conservation is a relatively new discipline which has developed as a response to biological issues such as extinction and marine habitats change.

Marine conservationists rely on a combination of scientific principles derived from marine biology, oceanography, and fisheries science, as well as on human factors such as, demand for marine resources and marine law, economics and policy, in order to determine how to best protect and conserve marine species and ecosystems. Marine conservation may be described as a sub-discipline of conservation biology.

Coral Reefs

Coral reefs are the epicenter of immense amounts of biodiversity and are a key player in the survival of entire ecosystems. They provide various marine animals with food, protection, and shelter which keep generations of species alive. Furthermore, coral reefs are an integral part of sustaining human life through serving as a food source (i.e., fish and mollusks) as well as a marine space for ecotourism which provides economic benefits. Also, humans are now conducting research regarding the use of corals as new potential sources for pharmaceuticals (i.e. steroids and anti-inflammatory drugs).

Unfortunately, because of the human impact on coral reefs, these ecosystems are becoming increasingly degraded and in need of conservation. The biggest threats include overfishing, destructive fishing practices, sedimentation, and pollution from land-based sources. This, in conjunction with increased carbon in oceans, coral bleaching, and diseases, means that there are no pristine reefs anywhere in the world. Up to 88% of coral reefs in Southeast Asia are now threatened, with 50% of those reefs at either "high" or "very high" risk of disappearing, which directly affects the biodiversity and survival of species dependent on coral.

This is especially harmful to island nations such as Samoa, Indonesia, and the Philippines, because many people there depend on the coral reef ecosystems to feed their families and to make a living. However, many fishermen are unable to catch as many fish as they used to, so they are increasingly using cyanide and dynamite in fishing, which further degrades the coral reef ecosystem. This perpetuation of bad habits simply leads to the further decline of coral reefs and therefore perpetuates the problem. One way of stopping this cycle is by educating the local community about why the conservation of marine spaces that include coral reefs is important.

Human Impact

Increasing human populations have resulted in increased human impact on ecosystems. Human activities has resulted in an increased extinction rate of species which has caused a major decrease in biological diversity of plants and animals in our environment. These impacts include increased pressure from fisheries including reef degradation and overfishing as well as pressure from the tourism industry which has increased over the past few years. The deterioration of coral reefs is mainly linked to human activities – 88% of reefs are threatened through various reasons as listed above, including excessive amounts of CO_2 (carbon dioxide) emissions. Oceans absorb approximately 1/3 of the CO_2 produced by humans, which has detrimental effects on the marine environment. The increasing levels of CO_2 in oceans change the seawater chemistry by decreasing the pH, which is known as ocean acidification.

Oil spills also impact marine environments, contributing to marine pollution as a result of human activity. The effects of oil on marine fish have been studied following major spills in the United States.

Techniques

Strategies and techniques for marine conservation tend to combine theoretical disciplines, such as population biology, with practical conservation strategies, such as setting up protected areas, as with marine protected areas (MPAs) or Voluntary Marine Conservation Areas. These protected areas may be established for a variety of reasons and aim to limit the impact of human activity. These protected areas operate differently which includes ares that have seasonal closures and/ or permanent closures as well as multiple levels of zoning that allow people to carryout different activities in separate areas; including, speed, no take and multi-use zones.

Other techniques include developing sustainable fisheries and restoring the populations of endangered species through artificial means.

Another focus of conservationists is on curtailing human activities that are detrimental to either marine ecosystems or species through policy, techniques such as fishing quotas, like those set up by the Northwest Atlantic Fisheries Organization, or laws such as those listed below. Recognizing the economics involved in human use of marine ecosystems is key, as is education of the public about conservation issues. This includes educating tourists that come to an area who might not be familiar with certain regulations regarding the marine habitat. One example of this is a project called Green Fins based in Southeast Asia that uses the scuba diving industry to educate the public. This project, implemented by UNEP, encourages scuba diving operators to educate their students about the importance of marine conservation and encourage them to dive in an environmentally friendly manner that does not damage coral reefs or associated marine ecosystems.

Technology and Halfway Technology

Marine conservation technologies are used to protect endangered and threatened marine organisms and/or habitat. These technologies are innovative and revolutionary because they reduce by-catch, increase the survivorship and health of marine life and habitat, and benefit fishermen who depend on the resources for profit. Examples of technologies include marine protected areas (MPAs), turtle excluder devices (TEDs), autonomous recording unit, pop-up satellite archival tag, and radio-frequency identification (RFID). Commercial practicality plays an important role in the success of marine conservation because it is necessary to cater to the needs of fishermen while also protecting marine life.

Pop-up satellite archival tag (PSAT or PAT) plays a vital role in marine conservation by providing marine biologists with an opportunity to study animals in their natural environments. These are used to track movements of (usually large, migratory) marine animals. A PSAT is an archival tag (or data logger) that is equipped with a means to transmit the collected data via satellite. Though the data are physically stored on the tag, its major advantage is that it does not have to be physically retrieved like an archival tag for the data to be available, making it a viable independent tool for animal behavior studies. These tags have been used to track movements of ocean sunfish, marlin, blue sharks, bluefin tuna, swordfish and sea turtles. Location, depth, temperature, and body movement data are used to answer questions about migratory patterns, seasonal feeding movements, daily habits, and survival after catch and release.

Turtle excluder devices (TEDs) remove a major threat to turtles in their marine environment. Many sea turtles are accidentally captured, injured or killed by fishing. In response to this threat the National Oceanic and Atmospheric Administration (NOAA) worked with the shrimp trawling industry to create the TEDs. By working with the industry they insured the commercial viability of the devices. A TED is a series of bars that is placed at the top or bottom of a trawl net, fitting the bars into the "neck" of the shrimp trawl and acting as a filter to ensure that only small animals may pass through. The shrimp will be caught but larger animals such as marine turtles that become caught by the trawler will be rejected by the filter function of the bars.

Similarly, halfway technologies work to increase the population of marine organisms. However, they do so without behavioral changes, and address the symptoms but not the cause of the declines. Examples of halfway technologies include hatcheries and fish ladders.

Extinct and Endangered Species

Marine Mammals

Baleen whales were predominantly hunted from 1600 through the mid-1900s, and were nearing extinction when a global ban on commercial whaling was put into effect in 1986 by the IWC (International Whaling Convention). The Atlantic gray whale, last sighted in 1740, is now extinct due to European and Native American Whaling. Since the 1960s the global population of monk seals has been rapidly declining. The Hawaiian and Mediterranean monk seals are considered to be one of the most endangered marine mammals on the planet, according to the NOAA. The last sighting of the Caribbean monk seal was in 1952, and it has now been confirmed extinct by the NOAA. The vaquita porpoise, discovered in 1958, has become the most endangered marine species. Over half the population has disappeared since 2012, leaving 100 left in 2014. The vaquita frequently drowns in fishing nets, which are used illegally in marine protected areas off the Gulf of Mexico.

Sea Turtles

In 2004, the Marine Turtle Specialist Group (MTSG), from the International Union for Conservation of Nature (IUCN), ran an assessment which determined that green turtles were globally endangered. Population decline in ocean basins is indicated through data collected by the MTSG that analyzes abundance and historical information on the species. This data examined the global population of green turtles at 32 nesting sites, and determined that over the last 100–150 years there has been a 48–65 percent decrease in the number of mature nesting females. The Kemp's ridley sea turtle population fell in 1947 when 33,000 nests, which accounted for 80 percent of the population, were collected and sold by villagers in Racho Nuevo, Mexico. In the early 1960s only

5,000 individuals were left, and between 1978 and 1991, 200 Kemp's Ridley Turtles nested annually. In 2015, the World Wildlife Fund and *National Geographic Magazine* named the Kemp's ridley the most endangered sea turtle in the world, with 1000 females nesting annually.

Fish

In 2014, the IUCN moved the Pacific bluefin tuna from "least concerned" to "vulnerable" on a scale that represents level of extinction risk. The Pacific bluefin tuna is targeted by the fishing industry mainly for its use in sushi. A stock assessment released in 2013 by the International Scientific Committee for Tuna and Tuna-Like Species in the North Pacific Ocean (ISC) shows that the Pacific bluefin tuna population dropped by 96 percent in the Pacific Ocean. According to the ISC assessment, 90 percent of the Pacific bluefin tuna caught are juveniles that have not reproduced.

Between 2011 and 2014, the European eel, Japanese eel, and American eel were put on the IUCN red list of endangered species. In 2015, the Environmental Agency concluded that the number of European eels has declined by 95 percent since 1990. An Environmental Agency officer, Andy Don, who has been researching eels for the past 20 years, said, "There is no doubt that there is a crisis. People have been reporting catching a kilo of glass eels this year when they would expect to catch 40 kilos. We have got to do something."

Marine Plants

Johnson's seagrass, a food source for the endangered green sea turtle, is the scarcest species in its genus. It reproduces asexually, which limits its ability to populate and colonize habitats. Data on this species is limited, but it is known that since the 1970s there has been a 50 percent decrease in abundance.

Marine Conservation

Modern marine conservation first became globally recognized in the 1970s after World War II in an era known as the "marine revolution". The United States federal legislation showed its support of marine conservation by institutionalizing protected areas and creating marine estuaries. In the mid-1970s the United States formed the International Union for Conservation of Nature (IUCN). Through this program, nations could communicate and make agreements about marine conservation. After the formation of the IUCN, new independent organizations known as non-governmental organizations started to appear. These organizations were self-governed and had individual goals for marine conservation. At the end of the 1970s, undersea explorations equipped with new technology such as computers were undertaken. During these explorations, fundamental principles of change were discovered in relation to marine ecosystems. Through this discovery, the interdependent nature of the ocean was revealed. This led to a change in the approach of marine conservation efforts, and a new emphasis was put on restoring systems within the environment, along with protecting biodiversity.

Overabundance

Overabundance occurs when the population of a certain species cannot be controlled. The domination of one species can create an imbalance in an ecosystem, which can lead to the demise of other species and of the habitat. Overabundance occurs predominately in invasive species. Cargo ships introduce new species into different environments through releasing ballast water into an ecosystem. A tank of ballast water is estimated to contain around 3,000 non-native species.

The San Francisco Bay is one of the places in the world that is the most impacted by foreign and invasive species. According to the Baykeeper organization, 97 percent of the organisms in the San Francisco Bay have been compromised by the 240 invasive species that have been brought into the ecosystem. Invasive species in the bay such as the Asian clam have changed the food web of the ecosystem by depleting populations of native species such as plankton. The Asian clam clogs pipes and obstructs the flow of water in electrical generating facilities. Their presence in the San Francisco Bay has cost the United States an estimated one billion dollars in damages.

References

- Threats-to-marine-environment-you-must-know, environment: marineinsight.com, Retrieved 23 August, 2019

- "Facts and figures on marine biodiversity | United Nations Educational, Scientific and Cultural Organization". Www.unesco.org. Retrieved 2018-12-01

- Overfishing, commerce, Topics: see-the-sea.org, Retrieved 24 January, 2019

- Derraik, José G.B (1 September 2002). "The pollution of the marine environment by plastic debris: a review". Marine Pollution Bulletin. 44 (9): 842–852. Doi:10.1016/S0025-326X(02)00220-5. ISSN 0025-326X

- Habitat-alteration, habitat, topics: see-the-sea.org, Retrieved 25 February, 2019

- US Department of Commerce, National Oceanic and Atmospheric Administration. "What is a lagoon?". Oceanservice.noaa.gov. Retrieved 2019-03-24

- Ritical-issues-marine-habitat-destruction, oceans, environment: nationalgeographic.com, Retrieved 26 March, 2019

- Jenkins, Lekeliad (2010). "Profile and influence of the successful fisher-Inventor of marine conservation technology". Conservation and Society. 8: 44. Doi:10.4103/0972-4923.62677

- Toxpol, toxic, pollution, pollution; see-the-sea.org, Retrieved 27 June, 2019

- "Background | Vulnerable Marine Ecosystems | Food and agriculture organisation of the united nations". www.fao.org. Retrieved 9 January 2018

Relationship of Humans and Oceans

5

- **The Ocean and Humans**
- **What is the Importance of Oceans in Human Life?**
- **Human Health and the Ocean**

Oceans are the lifeline for human survival as majority of oxygen is produced by the sea plants. They also play a vital role in improving the mental, psychological and emotional wellbeing of humans. The chapter closely examines the relation between human beings and the sea as well as the effect of ocean pollution on humans.

The Ocean and Humans

The full extent of this statement is key to any future resolution of today's challenges to the natural health and social value of the ocean. First, there is the affirmation of human presence — action and reaction — in all aspects of the natural world. Denial of the impact, positive or negative, is simply not a fact of life. Second, there is the finality of inextricability, the certainty that there can be no separation one from the other, no compromise of the actuality of connection. Third, there are the implications of the prefix, inter: to be located or existing between, in the midst, as in inter-grated; to be reciprocal or carried on between, as in inter-national; or to be occurring among, as in inter-vening. There are linguistic subtleties here that relate to nuance that, when amplified to a global scale, have incontrovertible meaning and significance.

How can we better communicate this connection? For example, most students learn about the water cycle in their earliest science classes. They see and understand the circular inter-action between ocean water, evaporation, circulation in the atmosphere, and condensation into fog or rain or snow far inland that further deposits and flows through run-off, streams, lakes, rivers, to an extent ending back again in the ocean near or far from each drop's point of origin. It is simple, elegant, easy to explain, and so most students retain it as a fundamental understanding of a natural system. But what about the human impacts of this circulation? While these may seem obvious, it is surprising how disconnected this knowledge is from understanding of the social consequences of the cycle as essential for our daily lives in the form of drinking water, irrigation, sanitation, manufacture, and so much else. When we claim that the ocean begins at the mountaintop and descends to the abyssal plain, we are amazed at the surprise such a declaration engenders, as if we have re-defined the ocean far beyond and in some original way from how it is conventionally understood as a distinctly separate place apart from the land.

Another similar example pertains to our patterns of consumption and exchange. Most people don't understand that almost every thing we make or purchase for our use has its economy and efficiency affected by maritime transportation and trade. Much of our energy, appliances, electronics, automobiles, processed foods, computers and communications, and even financial products such as currency and trading, are produced somewhere else and exchanged via ships or underwater cables that are, even in port cites such as New York or Shanghai, located away from the concentrated populations that consume these goods and services. When we interrupt this delivery, as a result of market forces, tariffs, regulations, or other economic or political actions, this global network slows or stops with further devastating inhibition of world security and stability. This ocean system is invisible and necessary as a structure for the circulation of goods that unites us in the best of times and separates and alienates us in the worst.

Finally, there are connections of people and ideas. Never have the people of this world been more mobile, moving as business executives, tourists, migrants and refugees seeking opportunity or fleeing tyranny. Never have ideas and innovations been more shared between teachers, students, policy-makers, governors, creators, and curious individuals who find connection though art, language, and invention. We have all become inter-connected citizens of the world through media and information facilitated by the same network of connection that brings us to the admixture of things and people that we call civilization.

Connection of to the Ocean

Oceans and Life Evolution

Changes in climate cycles from warm to droughts drove man's ancestors to search for fields with water sources leading to an evolutionary response making them turn into a human-like species. Ever since humans started exploring, water has been the guiding light, supporting both the discovery of new territories, races as well as the animal evolution like the story of the whales whose ancestors Pakicetus were once land mammals over 40 million years ago. Water is indeed tightly connected to the creation and evolution of life.

Oceans are a lifeline for human survival, in fact, water makes up 72% of the earth's surface and 70% of our body. Sea plants produce half of the world's oxygen and help to absorb one-third of human-caused gas emissions. Water plays vital functions in the human body such as digestion, absorption, circulation, the creation of saliva, transportation of nutrients, and the maintenance of body temperature.

Moreover, oceans regulate earth's system like the weather and climate that forms clouds that provide us with fresh water. Both, impact our everyday life, they shape our vegetation and soil, and define what food supplies grow as well as supply living and nonliving resources-minerals to renewable energy and marine biotechnology to support life as we know it.

Human-ocean Connection

Living near or having close contact with the sea improves the mental, emotional and psychological well-being. Oceans are a powerful source of energy that enables humans to disconnect from the hyperstimulated modern life, in this way we enter into a meditative state that renews the brain and body in order to feel refreshed, calmed, satisfied and present at the moment.

Besides, salt water has many health benefits as it enhances the immune system, boost circulation and hydrates the skin as well as relaxing the muscles and helping to reduce insomnia.

Nowadays technology is an integral part of life but there is still a lack of knowledge and activism regarding environmental issues. The Paralenz team believes in commitment, therefore, we have combined the two for the benefit of raising awareness. We care about the oceans that's why we have created the first underwater camera that allows everyone from children to professional divers to enjoy the immeasurable beauty of this powerful force of nature and so we can all start protecting it and giving anybody the chance to explore its magical underwater world.

What is the Importance of Oceans in Human Life?

We can make a long list of how the oceans and marine life are important to us. Oceans cover more than 70 per cent of the earth's surface. They contain 99 per cent of the living space on earth. Without this space for organisms to survive, there would be fewer phyla of animals on the earth. Perhaps this is the most important reason to protect the oceans to preserve the biodiversity of the Earth.

Biodiversity

Coral reefs, salt marshes, estuaries and mangroves and sea grass beds are just a few of the ocean environments which support a large number of different species of organisms, that is, have a high biodiversity. Estuaries are places where freshwater rivers and streams flow into the ocean, mixing with the seawater.

Along with coral reefs, estuaries sustain 75 per cent of commercial fish and shellfish during some point of their life cycles! Spawning organisms make reefs and estuaries their home because animals can find abundance of food and excellent protection from predators.

The fish, anemones, sea cucumbers and sea fans that populate the coral reefs all work together in symbiosis. Mangroves not only act as nurseries for commercially important marine species, they also act as a filtering system for coastal water. Sea grass beds, mangroves and coral reefs are crucial to providing protection against shoreline erosion and flooding.

The sandy shores are home to fiddler crabs and burrowing worms, as well as a feeding ground for birds. Without coral reefs and estuaries, our oceans would lose many, many organisms that are important to both humans and other marine life.

Natural Resources

The ocean floor habitat is not as well known as coral reefs or coastal areas, but it is very important to all the organisms that live on the bottom (benthic organisms), as well as commercially important. The continental shelves and ocean floor are home to many important minerals, including oil and natural gas. Natural gas and oil play a major role in meeting energy needs.

Climate and Weather

Did you know warm ocean waters provide the energy to fuel storm systems that provide fresh water vital to land-dwelling organisms? The oceans interact with and affect global weather and climate. As the air passes over warm waters, it rises due to warming. As it cools, condensation of the water creates rainfall. If the air passes over cooler waters, it cools and sinks. Air moves from high to low pressure areas.

Transportation

Not only are oceans important to sustain life, but also for moving materials that we use. Without commercial ships and barges, transportation of goods from place to place would be much more difficult and expensive. Cities which have good natural harbours always had an advantage, and even today are some of the largest cities in the world.

Importance of Ocean in the Water Cycle

The ocean plays an important role in the water cycle. The ocean holds about 97 percent of the total water on the Earth. About 78 percent of the global precipitation and 86 per cent of global evaporation takes place over the ocean.

Economy

The ocean is also vital to our economy. More than 66 percent of the world's population lives within 100 kilometres of the coastline. Many of the foods and products that we eat, or use as medicine contain ingredients from the sea. Carrageenan, a compound found in red algae, is used in peanut butter and toothpaste.

Compounds from ocean sponges and cartilage from sharks are being used in medication to help fight the battle against cancer. Great care is being taken in the research of marine-based drugs to prevent the depletion of important natural marine resources.

Human Health and the Ocean

The interaction between people and the ocean—particularly the coastal ocean—allows for instances in which the ocean and its inhabitants can have negative effects on human health. The greatest biological risk posed to people is from eating tainted seafood, although other factors also can harm humans: for example, the discharge into waterways of organic and inorganic wastes, including toxic substances; the global transport of microorganisms by shipping; and marine animals that bite or sting. Physical processes affecting human health include rough seas that threaten both small and large vessels; weather disturbances such as typhoons and hurricanes; rogue waves; and surges from storms and tsunamis.

On a positive note, the ocean may provide medical benefits: pharmaceutical companies collect aquatic organisms with the goal of creating new drugs that prevent or cure disease. Additionally, when people visit the ocean for tourism or recreation, there are positive effects on local economies and on human well-being. Even more significantly, marine phytoplankton conduct about half of the Earth's photosynthesis, producing oxygen and drawing down carbon dioxide. Phytoplankton form the basis of most oceanic food webs, thereby providing the food that is ultimately transferred up food chains to fish and shellfish.

Seafood and Harmful Algal Blooms

One cause of ocean-mediated illness in humans is harmful algae. Of the five thousand species of phytoplankton, about three hundred species can form "blooms" with concentrations high enough to color the water, creating the so-called "red tides" and "brown tides." At least ninety of the bloom-forming species are harmful to humans or animals, particularly through their ability to produce biotoxins.

When filter feeders are present during toxic algal blooms, they can concentrate algal toxins in their tissues. When people eat the fish or shell-fish containing these toxins, they may develop illnesses that affect the nervous system or the stomach and intestine. Roughly two thousand cases of human poisoning are recorded globally each year from eating fish and shell-fish contaminated with algal toxins, resulting in about three hundred deaths per year.

From the 1970s to present, there has been an apparent increase in the number, intensity, and global distribution of harmful algal blooms. This phenomenon is likely the result of multiple factors, including increased data reporting. For example, a heightened awareness among the public and scientific community has led to increased and more accurate monitoring of coastal algal blooms. The number of harmful algal blooms reported in the United States increased from two hundred per year in the 1970s to seven hundred per year in the 1990s.

Another hypothesis for the increase in harmful algal blooms stems from the overabundance of nutrients flowing into coastal waters. Varied sources contribute to the nutrient excess: discharges from sewage treatment plants; untreated sewage; runoff from farmland and overfertilized lawns; air emissions from factories and automobiles that ultimately are entrained in precipitation; loss of wetlands that take up nutrients; and poor erosion control. This human-induced overenrichment may favor harmful algal species that better use these nutrients than their non-toxic counterparts.

Inorganic and Organic Wastes

Metals (such as mercury) and chemicals (such as polychlorinated biphenyls) have been used in industrial processes and discharged into waterways over long timescales, from decades to centuries. Subsequently, some toxins have been incorporated into aquatic plants and animals.

Since the environmental movement coalesced in the 1970s, the industrial discharge of pollutants has received considerable attention, and many harmful pollutants have been reduced or eliminated from point-source discharges at factories. Nonetheless, some toxins have long residence times—persisting in the environment for decades—and remain at dangerously high levels in the food web and sediments. Consequently, unacceptably high concentrations of toxic substances are found in fish and shellfish tissues in many waterways. Local authorities thus set restrictions on harvesting and limits on consumption of tainted seafood, especially for children and pregnant women.

Food from the sea provides nourishment for human populations. Yet seafood contaminated with natural biotoxins, human-made chemicals, or sewage-related microbes poses a risk to human health. Developed countries generally have safeguards that minimize risks, whereas developing countries may struggle with inadequate public health resources.

Sewage

In countries without sewage treatment facilities, untreated sewage is released into coastal waters. Also, raw sewage may reach coastal waters when water runoff from big storms causes sewage treatment plants to bypass treatment. Finally, some human viruses and bacteria are resistant to chlorination at sewage treatment plants, and they are discharged with the treated effluent into nearby waterbodies.

Filter-feeding animals, such as oysters and clams, can concentrate pathogens, such as hepatitis A and Norwalk virus, found in sewage. The pathogens then can be transmitted to people when shellfish, especially if un-cooked, are eaten. Shellfish beds downstream from sewage treatment plants often are closed after storms until pathogens are naturally released from the shellfish. In a more straightforward route of infection, illness can also occur by direct contact with contaminated water containing viral, bacterial, or protozoan pathogens.

Microbes

Ships require ballast to control their stability and balance, and until the 1890s, rocks, dirt, and other forms of "dry ballast" were used. Today, ships use water for ballast by loading water aboard in one coastal port and discharging it at successive ports of call. Ballast water consequently contains everything that is in coastal water—viruses, bacteria, protozoa, phytoplankton, zooplankton, and fish. Some of these organisms can colonize new areas where the ballast water is discharged. hips can globally transport pathogenic microorganisms in ballast water. Many microorganisms have survival strategies, such as the formation of cysts, which enable them to withstand periods of inhospitable conditions. Cysts are an aquatic version of tree seeds that, in this specific case, allow microorganisms to survive confinement in darkened ballast-water tanks. Cysts of many species of toxic algae have been found in ballast water and in sediments in ballast-water tanks.

This diver comes face-to-face with a wolf eel, which can inflict a painful bite. Biting and stinging animals in the ocean can cause human injuries ranging from mild to severe, and in rare cases, life-threatening.

The bacterium that causes the human disease cholera can be transported in ballast water. Circumstantial evidence suggests that a previously unreported strain of cholera found in oysters and fish-gut contents in Mobile Bay, Alabama during 1991, was transported there by ballast water from ships that arrived from Latin America, which had an ongoing cholera epidemic. Although no cases of cholera were reported in Alabama, ballast-water samples collected later contained the epidemic-causing strain of cholera, and this incident highlighted the capability of ballast water to transport pathogens.

Animals

Humans can be harmed through direct contact with some marine animals. Although shark attacks receive the most attention, they are relatively rare compared to bites and stings from a variety of other animals, including sea snakes, eels, stingrays, jellyfish, and sea nettles.

Some of the most common stinging marine animals are found in the phylum Cnidaria, which includes the jellyfish, sea nettles, sea anemones, corals, and hydras. The Cnidarians' stinging cells are used to capture prey, and humans can inadvertently be stung as well. The stings from most species are only irritating, but a few animals, including the Portuguese man-of-war (Physalia physalis), can cause life-threatening reactions among people who are sensitive to the venom or have an allergic reaction to a sting.

Storms and Rough Seas

Also dangerous is the sea itself. Hundreds of people drown in the oceans every year while swimming, surfing, boating, or fishing. Coastal flooding caused by hurricanes and other events is an ever-present threat to human health and property.

One particularly dangerous aspect of ocean storms is the storm surge. This relatively sudden rise of normal sea level occurs when storm winds and tides combine to transport large volumes of water to coastal areas. Storm surges have been particularly lethal in low-lying river deltas, which often are heavily populated. For example, in 1900, a 6-meter (20-foot) storm surge and hurricane-force winds combined to kill at least 5,000 people in Galveston, Texas.

Changes in Ocean Currents and Climate

Ocean currents are a major engine for driving the Earth's climate, and changes in currents can alter climates throughout the world. El Niño is a particularly important disruption in the ocean–atmosphere system. During normal conditions, trade winds blow to the west across the tropical Pacific Ocean, piling up warm surface water in the western Pacific, and cold, deeper water rises up, or upwells, off the west coast of South America. This cold water is rich with nutrients, and it supports a phytoplankton community, which, in turn, is grazed by anchovies.

A slackening of the trade winds, called an El Niño event, results in warmer water on the western side of the Pacific basin spreading to the eastern side of the basin. Thus, off South America, cold water is not brought to the surface as efficiently by upwelling. This phenomenon can lead to a decreased catch of anchovies off the South American coast, flooding in Peru, increased rainfall across the southern part of the United States, and droughts in the southwestern Pacific. In addition to havoc caused by floods and mudslides, El Niño can cause secondary effects; for example, El Niño was implicated in increased cases of cholera in Peru in 1991 and 1998. Understanding and predicting El Niño phenomena is a major goal of climate researchers as well as public health authorities and disaster relief organizations worldwide.

Rough seas pose a danger to commercial ships, fishing boats, research ships, and other oceangoing vessels. This National Oceanic and Atmospheric Administration research vessel encounters huge waves in the Bering Sea.

With a view to the future, scientists predict that global temperatures could rise by 2 °C (3.6 °F) or more by the year 2100. With a temperature rise may come expansions in the ranges of marine organisms, leading to a possible increase in human illness. However, citizens and governments are increasingly aware of the impact that humans have on the global environment. One can hope that human populations become better custodians of the ocean and coastal waters in the future by, for example, reducing the flow of toxic substances and excess nutrients to coastal waters.

References

- Why-are-we-so-connected-to-the-ocean: paralenz.com, Retrieved 29 August, 2019

- What-is-the-importance-of-oceans-in-human-life, knowledge: shareyouressays.com, Retrieved 30 January, 2019

- Human-Health-and-the-Ocean, Ge-Hy: waterencyclopedia.com, Retrieved 1 February, 2019

- Ocean-pollution-and-effects-on-humans-and-marine-life, guides: sailsquare.com, Retrieved 3 March, 2019

PERMISSIONS

INDEX

www.ingramcontent.com/pod-product-compliance
Lightning Source LLC
Chambersburg PA
CBHW082016190326

41458CB00010B/3209